Conceptual PHYSICS

The High School Physics Program

Second Edition

Paul G. Hewitt

Teaching Guide

Paul G. Hewitt

Addison-Wesley Publishing Company

Menlo Park, California • Reading, Massachusetts • New York
Don Mills, Ontario • Wokingham, England • Amsterdam • Bonn
Paris • Milan • Madrid • Sydney • Singapore • Tokyo
Seoul • Taipei • Mexico City • San Juan

ACKNOWLEDGEMENTS

Many people have helped me with the ideas that make up this teaching guide. I am most grateful to Charlie Spiegel for providing insightful suggestions for every chapter of both the textbook and this *Teaching Guide*. Thanks also to resourceful Ronald E. Lindemann for many interesting examples and applications of good physics. I am also grateful to contributors Chase Ambler, Clarence Bakken, Bruce Eddman, Marshall Ellenstein, Robert L. Hare, Tucker Hiatt, Paul Hickman, John Hubisz, Kathy Wong Nirei, Bruce Ratcliffe, Paul Robinson, Nate Unterman, Nancy Watson, and Helen Yan.

Cover photograph © Ben Rose/The Image Bank

ISBN 0-201-28656-4

1 2 3 4 5 6 7 8 9 10-ML-95 94 93 92 91 (printed in 1991, ©1992)

Contents

Introduction

Conceptual Physics — The High School Physics Program is a way of teaching physics that should stimulate your students' higher-level cognitive skills. By contrast, physics courses that emphasize mathematical problem-solving are often "watered down" in order to serve students who lack mathematical skills. In *Conceptual Physics*, nothing is watered down with one exception — laborious algebraic problems are omitted.

Conceptual Physics features the three-stage learning cycle with emphasis on *doing physics*, as described on the next page. The rule of teaching here is **concepts before computation**.

The value of teaching physics conceptually is not in minimizing mathematics, but in maximizing the use of students' personal experience in the everyday world — in their everyday language. Students need not see physics as a hodgepodge of mechanistic equations or only as a classroom or laboratory activity. They need to see physics everywhere, as a part of everything they experience. People with a conceptual understanding of physics are more aware of their surroundings, just as a botanist taking a stroll through a wooded park is more aware of the trees and plants and the life that teems in them. Richness of life is not only seeing the world with one's eyes wide open, but also knowing what to look for.

What many students enjoy about a course is not finding that it is easy, but finding that they can comprehend difficult material. Students know the reputation of physics — that it is intellectually demanding. Because of this reputation and because we are teaching comprehensible physics (i.e., enjoyable physics), we find ourselves highly valued by our students. It's nice to be rewarded with a greater-than-usual number of appreciative students and with the personal satisfaction that comes with being appreciated for the best of reasons — helping students to get the best from their own brains.

Because physics has always been an excellent way to teach mathematics, it has been used for that purpose. But physics can serve a higher purpose — to teach students how to THINK! There are very few courses that teach thinking. *Conceptual Physics* is one of them.

How to Use *Conceptual Physics*

The Three-Stage Learning Cycle

Conceptual Physics is a program that utilizes the three-stage learning cycle, which was developed more than 20 years ago by Robert Karplus. Stage 1, **Exploration,** is done before students are introduced to material in the textbook. In this first stage, students do an activity in the *Conceptual Physics Laboratory Manual* authored by Paul Robinson (Student Edition code 28653; Teacher's Edition code 28654). In Stage 2, **Concept Development,** the focus is on the textbook and the *Concept-Development Practice Book* (Student Edition code 28658; Teacher's Edition code 28659). Stage 3, **Application,** focuses on experiments in the lab manual, follow-through questions and puzzles in the *Next-Time Questions Book* (code 28655), and Computational Problems in this *Teaching Guide*. This three-stage method of teaching optimizes learning and is explained below.

Stage 1: Exploration.

Before putting tires on your car you first jack up the car. Similarly with students, before putting ideas in their heads, you first get them interested. The Activities in the lab manual are designed to do this — to spark an interest that will make your treatment of material more meaningful. Information is best learned if it answers questions — that is, if it fills a need. The Activities set the stage for student's own questions. The Activities also ensure that all students have common experience from which to develop concepts. Thus it is very important that exploratory Activities precede your presentation. Some Activities may take only a few minutes, while some may take the better part of a whole class period. The Activities in the lab manual are distinguished from Experiments by the grey tab at the edge of the page. Lab Experiments (Stage 3) have solid black tabs.

Stage 2: Concept Development.

Concept development may take place via reading assignments, lectures and demonstrations, Concept Development Practice Page assignments, lab work, and/or class discussions. Lectures and class discussions should relate closely to chapter material. They should not be "off-the-wall" presentations or discussions about obscure material, nor should they be a verbatim presentation of material in the textbook. Successful classroom presentations fall between these extremes. This *Teaching Guide* contains suggested lectures and demonstrations that make available a selection of closely related ideas that are not in the textbook. They provide you with interesting ideas that supplement those in the textbook.

Stage 3: Application.

The follow-through to concept development is student *use* of the concepts. You can instigate this by leading a class discussion of worked-out answers to some of the Think and Explain exercises found in the text at the end of each chapter.

Conclude class periods by posing more complex questions for students to think about before the next class session. These are the *Conceptual Physics Next-Time Questions*, which are also available as part of your teacher resource materials. These provide a continuity device, an interesting way of picking up where you left off. Your next class can begin by discussing the answer. Post these on your physics bulletin board, or make transparencies of them for overhead projection.

If you treat computational problem solving beyond numerical illustrations, please let this be in the application stage — *after* students have demonstrated an understanding of the physics concepts. Such problems are found not in the textbook, but are in this *Teaching Guide*. The primary difference between a conceptual and a computational course is the emphasis on the exploratory and concept-development stages before applying the concepts with computational problem solving.

Laboratory Experiments (as noted by the black tab at the edge of the page in the lab manual) comprise the strongest component of the application stage.

Lectures

Lecturing skill is in the sequencing of ideas; the understanding that one idea leads to the next. I suggest the policy of asking **check questions.** After discussing an idea and before developing it further or advancing to a new idea, give your class a question to consider. I do this as follows: "If you understand this — if you really do — then you can answer the following question." Then I pose the question slowly and clearly, usually in multiple-choice form for a short answer, and ask the class to respond — usually in writing. Next, I ask them to check their response with a neighbor. A few check questions during lecture brings students into an active role. The check question procedure may also be used to *introduce* ideas. A discussion of the question, the answer(s), and some of the associated misconceptions gets more attention than the same idea presented as a statement of fact.

Equations are important in any physics course. In a conceptual physics course an equation is not a recipe for plugging in numerical values, but rather it is a **guide to thinking.** The equation tells your students what variables must be considered in treating an idea. How much an object accelerates, for example, depends not only upon the net force on it, but also on its mass. Consideration of the equation $a = F/m$ reminds students of this. Does gravitation depend on an object's speed? Consideration of $F = GmM/d^2$ shows that it doesn't.

Please do not get into "second or higher-order effects" in this introductory course — such as introducing the idea that mass depends on speed before special relativity has been covered. Nothing inhibits learning more than information overload! To cover every possible nuance may impress some of your colleagues and even a few of your brighter students, but the extra depth of your plow setting will probably dampen learning for most of your students.

We all acknowledge the value of **demonstrations.** Watching them and doing them, however, are entirely different. Performing demonstrations is an art. It is one of those activities that looks easy from the sidelines but is demanding in practice. Perhaps that's why few teachers feature demonstrations in their teaching, although when new to the profession they strongly intend to bring life to their classes with wonderful demonstrations. But demonstrations are work! Performing them is fun, but getting the necessary equipment in order is not always easy. Demonstrations that require simple equipment are listed in the suggested lectures. Please make the effort to include some demonstrations in every lecture. If you discuss falling objects, at least drop your chalk or an eraser on the table! Remember the Chinese proverb, "I hear, and I forget; I see, and I remember ...". Let your students *see* you doing physics while you lecture.

The complete proverb is, "I hear, and I forget; I see, and I remember; *I do, and I understand.*"

Practice Pages from the *Concept-Development Practice Book* are provided for every chapter and should be used to get your students *doing* physics. The Practice Pages are very different than the traditional worksheets of other courses. Working through the Practice Pages is an alternative to working through the algebraic problems that characterize the "plug-and-chug" type physics course. Practice Pages can generally be assigned right after your lecture and before homework is done and discussed. The Practice Pages can be completed by student pairs, in which case the person doing the writing should sign it and the person approving should sign or initial it (as is done in R&D reports for industry). Alternatively, the sheets can be done individually. Your students will find them stimulating, interesting, and fun.

The suggested lectures in this book are as the name implies, *suggestions*. They suggest a sequence of ideas that has worked for me and for others, and that may work for you. They offer many interesting tidbits not found in the textbook that can enliven your presentation. Some of the annotated suggestions in the margins of the textbook are included in these lectures, and some are not. Pick and choose the ideas that suit your style of teaching. What works for one teacher may or may not work for another. We all have to develop our own style of teaching. The suggested lectures in this book and the videotapes provide a sample base for developing your own noncomputational way of teaching.

Whatever your style, I strongly recommend that you use a lesson plan or some form of lecture notes. After teaching conceptual physics for 20 years, I still bring note sheets to every lecture. These sheets are simply a check list that I glance at from time to time in my lecture to be sure that I cover the ideas intended. If I don't cover all the ideas on the sheet, a mark or two will let me know next time what I missed or where I stopped.

Laboratory Work

Lab work is a very important part of this program. In addition to reinforcing concepts, lab experiments familiarize your students with methods of taking, recording, and analyzing data.

Some labs in the the *Conceptual Physics Laboratory Manual* call for use of a computer and instrument interfacing. These labs are optional. They were designed to be compatible with the *Apple II Series* computer, because the great majority of quality programs for science education have been written for the Apple. You will have ample lab experiments to choose from, both with and without the use of a computer. Teacher information about the labs is in the front of the *Laboratory Manual TE.*

Homework

You are the best judge of how much homework to assign, for only you know the level of your class. No homework at all leaves students dangling between classes, a little keeps them engaged, more keeps them busy, and too much blows them away. I recommend you devote one class period per chapter to guide student discussion of the homework that is handed in.

Answering the Review Questions from the text should be an achievable task for even your less-than-average students. Answers can be located in the text at the designated sections. This task ensures that the main points of the chapter have been covered. It is therefore reasonable to assign half or even all the Review Questions as homework. Answering the Think and Explain questions at the end of each chapter is more demanding, and some questions may stump even your best students. Pick and choose among the offerings, and assign them with discretion. Additional questions not in the text are included with each suggested lecture. Practice Pages from the *Concept-Development Practice Book*, although primarily intended for classroom activity, may be used as homework assignments. In assigning homework, recall your own experience as a student with courses that demanded so much work that you couldn't see the forest for the trees. Please do not overwhelm your students with excessive written assignments! You will have a handle on how much homework is right for your class from the feedback you get in the student discussion session.

In response to students who find physics difficult, be sure to remind them from time to time throughout the course that it *is* difficult, and not necessarily a sign of their own deficiencies. Remind them that they are coming to grips in a matter of days with ideas that took centuries to develop.

Testing

This program has two testing options. *Conceptual Physics Tests (code 28506)* is a book of blackline master tests that contains one test for each chapter plus two semester tests. Questions in the test booklet are in multiple-choice, true-false, and essay form.

A second option is *Addison-Wesley Test File, An Assessment Program (code 28505)*. It includes a test bank of over 1000 questions in multiple-choice, true-false, and essay form. You can choose specific questions from the test bank or use the two tests per chapter that have already been prepared for your convenience.

Software that produces tests with the Test File questions will be available. Contact your Addison-Wesley sales representative for complete information on this software.

There are two levels of test questions in the test bank. The first-level questions are comparable to the Review Questions found in the text. Tests from these questions should be taken with closed books. The second-level questions are more demanding. To discourage the notion that education consists of memorization and to encourage thinking, tests from these questions can be given with open books and/or open notes.

Selected References

Textbooks: *Conceptual Physics* was first published in 1971 by the College Division of Little, Brown and Company, Boston, Mass. The sixth edition, published in 1989 by Scott, Foresman Publishing Company and since then by Harper Collins differs from the high school text in several ways. In addition to having a generally higher reading level, the sixth edition has a chapter on general relativity, and it has a greater number of Think-and-Explain questions (called Exercises). Some of these may be useful to you in supplementing the chapter-end material. The seventh edition of the college version of *Conceptual Physics* is scheduled for 1994.

Another college textbook that features the conceptual approach to physics is *Physics — An Introduction, 2nd Edition*, by Jay Boleman, published in 1989 by Prentice Hall. This book nicely complements *Conceptual Physics*.

I highly recommend *Physics for the Inquiring Mind*, by Eric Rogers, published in 1960 by Princeton University Press. This book has been an inspiration to me, both as a teacher and an author. (It is a very thick book, so much so that my department chairman forbade me to adopt it for my "Descriptive Physics" course. Had I been allowed to share it with my students I might never have written *Conceptual Physics*.)

At a more difficult level are the three volumes of *The Feynman Lectures*, which are edited transcriptions of Feynman's lectures given to Cal-Tech students in 1961. These volumes, published by Addison-Wesley, were early inspirations that broadened my perspectives of physics. I recommend them to all physics teachers.

Magazines: The reading I recommend most highly is not a textbook, but rather the magazine of the American Association of Physics Teachers, *The Physics Teacher*. This monthly publication is a must for all physics teachers. It contains a wealth of information relating to high-school physics teaching. AAPT, College Park, MD 20742.

Audiotapes. *Moments of Discovery* by Arthur Eisenkraft, physics teacher at Fox Lane High School in Bedford, NY, is a valuable reference. Part I, which is a history of the discovery of fission, uses the actual recordings of the physicists involved with this achievement, including J. J. Thompson, Rutherford, Bohr, Fermi, Einstein, and others. Part II is a history of the discovery of the optical pulsar. This tape includes an impromptu recording of Cocke and Disney on the night they first observed the pulsar. It is the only known live recording of a scientific discovery as it actually took place. A teacher's guide with supporting material is included. The package was funded by the National Science Foundation, The Heideman Foundation, and The Friends of the Center for the History of Physics. The entire package is available for $85 from The Center for the History of Physics of the American Institute, 335 East 45 Street, New York, NY 10017.

Computer Software. The computer serves a variety of functions for classroom and laboratory use. Its optional role in the lab is described in the front matter of the *Conceptual Physics Laboratory Manual*.

Companion software to *Conceptual Physics* includes a *Laboratory Interfacing Disk* for Apple II series computers that enable the collection, recording, and graphing of data quickly and accurately. An interface box makes the game port easily accessible for the insertion of temperature probes (thermisters) and light probes (phototransistors). Another disk, *Conceptual Physics Laboratory Simulations I*, complements some of the labs in the lab manual. Still another disk, *Good Stuff!*, has programs of ultra-fast machine language graphics animation of physics concepts by Professor Robert H. Good of California State University, Hayward. These materials are available from Laserpoint, 1328 West Palo Alto Avenue, Fresno, CA 93711, phone: (209) 435-5273.

Some of the best software for the Apple II, IBM-compatible, and Macintosh computers is described in the lab manual and has been developed by David Vernier. For more information, contact Vernier Software, 2920 S.W. 89th Street, Portland, OR 97225, phone: (503) 297-5317.

For color IBM computers or compatibles, check out the *Physics Discovery Series*, developed by Roy Unruh of the University of Northern Iowa. These programs employ the same learning strategy used in the PRISMS material embraced in the lab manual. These programs are available from your local IBM software representative. For more information, contact IBM Corporation, PO Box 138-W, Boca Raton, FL 33429.

Videotapes. In 1983 my ex-student and long-time friend Dave Vasquez teamed up with Craig Dawson to videotape my classes at City College of San Francisco. From this footage of my lectures Addison-Wesley published a series of 12 tapes to accompany the first edition of *Conceptual Physics*. At this writing more than twice as many new tapes are being prepared of my Conceptual Physics class taught while on a two-year leave at the University of Hawaii at Manoa.

The usefulness of the tapes is variable. They introduce your class to the author of their book (which may result in greater interest in studying the textbook). They bring a guest lecturer into your class. They can be shown by a substitute whenever you can't make it to class. The tapes can also be used in your learning resource center as supplementary material, or loaned to students to view at home. Alternately, you can treat them as part of this teaching guide and view them privately to get ideas for your own lectures.

Still available from the 1984 City College footage is *Teaching Conceptual Physics* (60 minutes), *Classroom Demonstrations in Conceptual Physics* (60 minutes), and *Fusion Torch and Ripe Tomatoes* (45 minutes). The first tape is narrated highlights of the course that is intended for teachers, the second is non-narrated highlights of demonstrations of the course, and the third is my opening lecture that is designed to elicit interest in the course. For more information, contact ABC Video Services, 1140 Irving Street, San Francisco, CA 94122.

Another set of physics videotapes is the set of 12 high school adaptations of *The Mechanical Universe*. The series was created initially for a college audience and was produced at the California Institute of Technology. These tapes feature a historical perspective and have outstanding graphics. They are available from the Southern California Consortium, 5400 Orange Avenue, Suite 109, Cypress, CA 90630.

Videodiscs. The complete college course of *The Mechanical Universe* is available on videodisc. Call 800 - LEARNER. *The Puzzle of the Tacoma Narrows Bridge Collapse*, by Robert Fuller, Dean Zollman, and Tom Campbell, and *Physics and Automobile Collisions*, by Zollman and Fuller, are published by John Wiley & Sons, Inc., 605 Third Ave., New York, NY 10158. Other attention grabbers by Fuller and Zollman are *Studies in Motion*, which uses gymnasts, divers, and a dancer, and *Energy Transformations Featuring the Bicycle*. These are published by Great Plains Television Library, University of Nebraska, Box 80669, Lincoln, NE 68501. Charles Eames' popular *Powers of Ten* and *Tops* are available from videodisc educational distributors such as Itek Co., Inc. (800) 247-1603, free catalog; Videodiscovery, Inc., (800) 548-3472, free catalog; Optical Data Corp. (800)-524-2481; and Encyclopedia Brittanica Educational Corp., (800) 554-9862. The American Association of Physics Teachers has produced a set of six classics discs that include *PSSC* and *Project Physics* films (1991).

Filmstrips and Slides. Spectacular slides, most from Bob Greenler's *Rainbows, Halos, and Glories* (Cambridge University Press, 1980) are available from Mallmann Scientific Co., 20250 Jeffers Drive, New Berlin, WI 53151. Two sets of five-lesson filmstrips with audio tape are *Gravitation* by Lester Paldy, and *Optics* by

Jacqueline Spears — available from Prentice Hall Media, Englewood Cliffs, NJ 07632. If you are interested in a wider selection of slides and filmstrips, write for the free *AAPT Products Catalog* from the American Association of Physics Teachers, 5112 Berwyn Road, College Park, MD 20740-4100.

Further Information

For further information on all components of the *Conceptual Physics High School Program*, call Addison-Wesley's toll-free customer service number: (800) 447-2226.

Course Planning

The *Conceptual Physics Program* is based on a 170- to 180-day teaching year. I assume that most users will begin their course with Chapter 1 in the text and more or less follow sequentially through the book toward Chapter 40. It would be unusual, however, to cover every chapter in the book in one teaching year, given the distractions that usually occur. I take the view that the 40 chapters offer a wide selection for a 25- to 35-chapter course. All the units and chapters are important, which is why they are in the book, but when there isn't time to sufficiently cover them all, something has to give. It is up to the teacher to pick and choose course coverage.

The simplest way to limit course coverage is to simply omit an entire unit. I strongly recommend you **not** omit Unit I, *Mechanics*, because its physics is basic in the course. Although Units II and II are excellent material, omitting either of them helps trim content to allow for an unhurried coverage of the rest of the material. Unit IV is perhaps of highest initial interest to your students, and for that reason may not be a good candidate for omission. Unit V on electricity poses a dilemma, for although it is more abstract and more difficult than other units, it is likely to be the only formal exposure that your nonscience students will ever have to this subject — even though every facet of their lives is touched by electrical phenomena. For your science students, its conceptual nature as an introduction is very valuable. Unit VI, *Atomic and Nuclear Physics*, poses a similar dilemma. You will have to be the judge of your course content.

A major strength of the text, I feel, is that most chapters can stand on their own. Consequently, the omission of chapters within units still permits an adequate flow of the remaining material. This allows you to cover the concepts in every unit, in lieu of omitting entire units. Suggestions for omitting chapters with possible consequences are found in the notes that follow.

I suggest omitting Chapters 15 and 16 on special relativity. These can be omitted without detriment to the remaining chapters, even though the mass-energy relationship is again treated in Chapter 40. Other chapter candidates for deletion are Chapters 6: *Vectors*, 24: *Thermodynamics*, 30: *Lenses*, 31: *Diffraction and Interference*, and 38: *The Atom and the Quantum*.

If you try to cover 40 chapters in 36 weeks, it's easy to see you'll be covering 2 chapters a week some of the time. (Who gets a full 36 weeks for course content?) That doesn't leave much time for exploratory activities, experiments, and application activities — not to mention testing. To ease the crunch you might consider assigning some chapters as outside reading, without lectures, homework, and testing. This allows more class time for selected chapters. Unless you have a very good class and excellent teaching conditions, I suggest you not try to cover completely every chapter in the book.

Suppose, for example, that for the first half year you cover Units I and II and assign the relativity chapters as outside reading with no homework or testing. This gives you 18 weeks to cover 18 chapters (a bit more considering that Chapter 1 should take at most 2 or 3 days). One chapter per week may allow you enough time for activities (which often require less than a full class period), lecture, workbook exercises (again often less than a class period), homework discussion time, and an experiment. Chapters such as 3, 5, 10, and 14 take less time than the others. This allows some time for testing. If you want quizzes and exams to be a positive reinforcement tool, then give small quizzes often rather than only large comprehensive exams. Entry level students have difficulty with long-range retention of physics concepts.

This plan would leave you the second 18 weeks during which to cover 20 chapters. Again, you might consider assigning some chapters as outside reading, perhaps Chapters 24, 30, 33, and 38. This leaves you with a chapter per week and two weeks for spring fever. In this way, you may cover the entire book.

The notes following this section suggest lecture time for all the chapters of the book. The word *lecture* is more appropriate in college than it is in high school, and perhaps I should call these *instructional periods*. My personal experience has been with 50-minute lectures to large groups, so the material that I am used to covering in one lecture may be very different than what you will cover in one class period. After you compare the time you usually take to cover material to what I consider one or two lectures, you can translate my coverage time into your own. My pace is fairly brisk, because I treat the course as an overview of serious physics and I don't get caught up in having my students master all the details we encounter.

The lecture time per chapter breakdown that follows does not take into account class time for activities, experiments, workbook exercises, homework discussions, and testing. It is an estimate of relative lecture times for various chapters that may help you select the combination of lessons that best meets your class needs.

Unit I: Mechanics

The time spent on Chapter 2 can be long or short, depending on how much you expect to cover and how deep you wish to set your plow. I suggest you cover the rudiments of the chapter in two lectures; one on the distinction between speed, velocity, and acceleration, and the other on the physics of falling bodies. Next move on to Newton's laws — Chapters 3, 4, and 5. Chapters 3 and 5 can each be covered in a single lecture, and Chapter 4 will require at least two lectures. Chapter 6 on vectors is more involved, and if you don't go into detail, three lectures with added time for the workbook exercises should be adequate. The conceptual treatment of vectors is a useful foundation that goes beyond mechanics and is important. However, if this chapter is omitted from your course, then you should also omit Section 7.6 of Chapter 7. Momentum (Chapter 7) can be covered in two lectures, whereas three lectures may be required to cover all of energy in Chapter 8. Chapter 10 on center of gravity is a "shorty" and can be covered in one lecture. Gravitation is divided between Chapters 12 and 13, which can be covered in two or three lectures. Chapters 9 and 14, which cover circular motion and satellite motion, can each be completed in one lecture. Rotational mechanics, Chapter 11, will need at least two lectures. Chapters 15 and 16, on special relativity, require at least 3 lectures, when covered together. In order to shorten Unit I, these two chapters are the best candidates for omission. One reason is that they are abstract and remote from the students' everyday experience; another reason is their high level of difficulty.

Unit II: Properties of Matter

Even the briefest treatment of this unit should include Chapter 17 on atoms, which is background for much of the rest of the book. Chapter 17 can be covered in one and a half lectures. Chapters 18, 19, and 20 can each be covered in two lectures and are not prerequisites for the chapters that follow. Unit II, with the exception of some mention of kinetic energy in Section 20.7, may be taught before, or without, Unit I. With minor difficulties, Unit II can stand alone without Unit I. Many teachers omit Unit II when their students have covered this material in a chemistry or other science course.

Unit III: Heat

Knowledge of kinetic energy, potential energy, and the conservation of energy is a prerequisite for this unit. Chapters 21 and 22 can be covered in one or one and a half lectures each, and Chapter 23 (*Change of State*) and Chapter 24 (*Thermodynamics*) will each require two lectures. If time is tight, the best chapter to skim, or omit, is Chapter 24. The concepts of heat and thermal equilibrium (Chapter 21) make a useful background to similar concepts in electricity (Chapters 32 through 35), radiation (Chapter 23), and electromagnetic waves (Chapter 27). Although portions of this material make good background for some of the material to follow, it is not a prerequisite for following chapters.

Unit IV: Sound and Light

Each of these chapters is packed with information, and each requires at least two lectures for adequate coverage. Units I, II, and III are not prerequisites for this unit, so consider beginning your course with Unit IV. A brief treatment of this unit can omit Chapters 30 and 31. Chapter 30 is background for laboratory work, and if no lab work is done, it should be minimized or omitted. Section 28.11 on atomic spectra is useful background for the quantum physics covered in Chapter 38.

Unit V: Electricity and Magnetism

Much of Unit I as well as Chapter 25 of Unit IV are prerequisites for this unit, which is the most conceptually difficult in the book. Chapter 32 requires one and a half lectures, Chapter 33 requires one lecture, Chapters 34 and 35 each require two lectures, one lecture is needed for Chapter 36, and two lectures are needed for Chapter 37. Because the material is abstract, more time should be spent on the supporting activities in this unit. This unit is too difficult to be taught during the first half of your school year. For a light treatment of physics, it can be omitted entirely. However, the cost of its omission is rather high, considering the importance of electricity in the modern world and the remote likelihood of students encountering a serious study of it elsewhere.

Unit VI: Atomic and Nuclear Physics

Chapter 17 on atoms is a prerequisite for this unit. If you are not covering Unit II, consider doing Chapter 17 immediately before this unit.

Unit VI can then stand on its own. Chapter 38 can be covered in one lecture, and the following two chapters need two lectures each. For a short treatment, Chapter 38 can be assigned as reading or omitted entirely.

Appendixes

Graphical analysis is the topic of Appendix C. This may supplement motion analysis in Chapter 2 and much of the lab manual. More about vectors is in Appendix D, which can follow Chapter 6 for those who wish a full treatment of vectors. Exponential growth and doubling time, which doesn't fit well in any of the chapters of the book, is treated in Appendix E. It's not hard-core physics, but it is important enough for inclusion. If you have a stray period in your schedule, it will make a very interesting lecture.

1 About Science

Chapter 1 Planning Guide

• *The bulleted items are key: Be sure to do them!*

Topic	Exploration	Concept Development	Application
Scientific Method	• Act 1 (<1 period)	• Text 1.1–1.7/Lecture Con Dev Pract Pg 1-1	• Nx-Time Q 1-1

Video: *Introduction to Conceptual Physics*
Evaluation: Chapter 1 Test

See the Chapter Notes for alternative ways to use all these resources.

Objectives

After studying Chapter 1, students will be able to:
• Explain why physics is the basic science.
• Outline the five steps of the scientific method.
• Distinguish between an observation and a fact.
• Distinguish between a fact and a hypothesis.
• Distinguish between a hypothesis and a law or principle.
• Describe the circumstances under which a hypothesis or law must be changed or abandoned.
• Distinguish between the everyday meaning and the scientific meaning of theory.
• Explain why the refinement of theories is a strength in science.
• Distinguish between a hypothesis that is scientific and one that is not.
• Distinguish between science and technology.

Possible Misconceptions to Correct

• Physics is the most difficult of the sciences.
• Physics is applied mathematics.
• Facts are unchangeable.
• A theory is a scientific word for guess.
• It is bad science to change your mind.
• Science and technology are the same.
• There is no distinction between science and technology or between a hypothesis and a theory.

Introduction

I suggest you spend only one class period lecturing on Chapter 1. Although this material is important groundwork for science in general, it should not be considered material to *study*, like the following chapters are. It is enough for your students to be familiar with the objectives above. Save deeper digging for later. Then your course can start with hard-core physics without undue pressure.

The "Two-Word Game"

Here is a game you can play with your class; the "Two-Word Game." It illustrates physics as an activity to search for laws to explain, simplify, and predict. Tell your class you will say two words that have a particular relationship between them. You will then continue with other pairs of words that have the same relationship. (The relationship is the same for all the pairs of words.) The students are to figure out the relationship, just as scientists try to discover relationships in nature. When the students think they have figured it out, they then "predict" by giving two new examples **without saying the rule**. To make sure that they just didn't stumble onto examples that are correct, have them give additional examples. They are never to say the rule, just give examples that use the rule.

When they understand the rules of the game, start. It's *off* but not *on*. It's *cool* but not *cold*. It's *kitten* but not *cat*. It's *wall* but not *ceiling*. It's *Milwaukee* but not *Chicago*. It's *Illinois* but not *Wisconsin*. It's *nineteen* but not *twenty*. It's *three* but not *two*. It's *pepper* but not *salt*. It's *book* but not *text*. It's *tall* but not *short*. It's *knee* but not *elbow*.

The rule has to do with the spelling of the words, rather than any meaning they might convey. That is, a double letter appears in the first word and not in the second. If a student guesses, for example, it's *calf* but not *cow*, say it's neither *calf* nor *cow*. If the student should guess, its *sixteen* but not *seventeen*, say that it is both *sixteen* and *seventeen*.

Soon some students should figure out the rule. When some have, stop and ask how those who don't understand the game feel — confused? frustrated? stupid? That is the way humankind felt before scientists finally figured out some of nature's rules or laws.

Point out the parallel between this game and learning physics. If the student knows the rules, he or she doesn't have to memorize scores of examples and meaningless data. If the student knows the rules when called upon in class or a test, he or she could come up with examples that are new and original.

As is stated in the "To the Student" page of the textbook, learning the rules of the physical world is what physics is about. To know the rules allows one to see the world with meaning, and with more interest. A person with a knowledge of music better appreciates music, a person trained in art better appreciates art, and a person acquainted with the rules of nature better appreciates nature and is more alive to the beauty of the surrounding world. Do you agree that for every case where ignorance is bliss, there are far more cases where knowledge is even more blissful?

Photo Scrapbook

A course-long assignment to consider is having each student take a photograph of something that illustrates physics. Each photo should be an example of physics in everyday life. A title and a short paragraph should explain the photo. One example might be of a book lying on a table, with the title, "Sum of the forces equals zero" and a short explanation that the force of gravity on the book is balanced by the equal force of the table pushing up on the book. Another example, one that is omnipresent, is of the student who takes a picture of himself/herself in the mirror. The photo might be titled "Reflection," with a short paragraph on the law of reflection or on virtual images. As photographs are submitted, they can be displayed in a photo scrapbook or on a bulletin board. Students who initially find difficulty in choosing photo subjects finally come to the realization that a photograph of anything at all involves physics. The realization (not to be stated at the outset!) is the value of this activity. Physics is truly all around us!

The double-letter game and assignment of the photo scrapbook on the first day of class should be a good start. If you choose not to do these, discussion of them is still of some value.

With the popularity of video cameras these days, you might consider giving your students the option of doing a short video in place of the photograph.

Teaching Suggestions

This section describes some points that are not in the book but are related to particular sections of Chapter 1.

Section 1.1: Consider physics, chemistry, and biology, which is the simplest? Which is the most complex? Does the most complex science necessarily make for the most difficult course of study? Is it the "depth of the plow" or the field of study that makes a course difficult? In terms of science courses in school, physics has traditionally been more demanding of students than "easier" chemistry and biology courses are. Why? Simply put, understanding of the subject is expected more in physics than it is in the more complex chemistry and biology courses. Because physics is now taught conceptually, science students can logically begin with the study of physics, then take chemistry, and then progress to a serious course in biology. Conceptual physics, without the mathematical "roadblock" is comprehensible to nonscience students also.

Section 1.2: Although mathematics is the language of physics for physicists, it is not the focus of this course. The principal focus of this course will be on understanding physics visually. This course stresses comprehension and, except for elementary computations in the Practice Pages and lab part of the course, emphasis on computation is left for those who choose a follow-up course. With a base of conceptual comprehension first, computation later will be more meaningful.

Section 1.3: Prediction in science is different from prediction in other areas. In the everyday sense, one speaks of predicting what has not yet occurred, like whether or not it will rain on next Saturday.

In science, however, prediction is not so much about what *will* happen but about what *is* happening and is not yet noticed, like what the angular momentum is for a particular elementary particle — or new-found star. A scientist predicts what can and cannot happen, rather than what will or will not happen.

Section 1.4: It is important that a scientist be open-minded *and* skeptical. It is not enough to be skeptical, and it is not enough to be open-minded. Strive to be both.

Section 1.5: Most people think only about the correctness of their ideas and what that correctness means. A good scientist goes further and thinks about the possible wrongness of his/her ideas and what the consequences are if those ideas are wrong. Do you?

Section 1.6: Science is finding things out; technology is doing them. Both science and technology in themselves are neither good nor bad. What people or governments choose to do with them can be good or bad. It is the responsibility of all of us to see that science and technology are wisely used to promote the general well-being. Responsibility and authority should go together. Do they?

Section 1.7: One of the important objectives of this course is to help students think critically, which begins by making distinctions between things. When distinctions such as those between science and technology, hypothesis and theory, and force and pressure are understood, clearer thinking can follow.

Chapter 1 End-matter: I have no suggested additions to Chapter 1 end-matter, and urge that you move quickly into Chapter 2.

2 Motion

To use this planning guide work from left to right and top to bottom.

Chapter 2 Planning Guide
• *The bulleted items are key: Be sure to do them!*

Topic	Exploration	Concept Development	Application
Speed	Act 2 (<1 period) • Act 3 (<1 period)	• Text 2.1–2.3/Lecture Demo 2-1 Con Dev Pract Pg 2-1, 2.2	
Acceleration	• Act 5 (1 period) Act 6 (1 period)	• Text 2.4–2.8/Lecture	Exp 4 (>1 period) Nx-Time Q 2-1

Video: *Motion*
Evaluation: Chapter 2 Test

See the Chapter Notes for alternative ways to use all these resources.

Objectives

After studying Chapter 2, students will be able to :
• Explain the idea that motion is relative.
• Define speed and give examples of units for speed.
• Distinguish between instantaneous speed and average speed.
• Distinguish between speed and velocity.
• Describe how to tell whether or not a velocity is changing.
• Define acceleration and give examples of units for acceleration.
• Describe the motion of an object in free fall from rest.
• Describe the motion of an object thrown straight up until it returns and hits the ground, when air resistance is negligible.
• Determine the speed and the distance an object falls at any time after it is dropped from rest, when air resistance is negligible.
• Describe how air resistance affects the motion of falling objects.
• Explain why acceleration is a *rate of a rate*.

Possible Misconceptions to Correct

• Speed and velocity are the same.
• Acceleration is simply a change in velocity.
• How fast something goes when moving is how far it goes.

Demonstration Equipment

• [2-1] Textbook and two ordinary half-sheets of notebook or copy paper

Introduction

I have had mixed feelings about beginning the course with mechanics, and in particular with motion. This is because the study of motion can be quite demanding and calls for our students' highest cognitive skills. Motion is basic to the other areas of physics, however. Rather than beat around the bush and begin with less demanding material, we jump into the foundations of mechanics right away. As was stated earlier, the demand a course places on students has more to do with how deep the plow is set than it does with the field being plowed.

Therefore, Chapter 2 covers the essentials of kinematics using only simple examples. Only motion in a straight line is treated. Only uniform acceleration with emphasis on falling objects is discussed. Make the distinction between velocity and acceleration with the clear understanding that freely-falling objects fall in a predictable way, and then move on to Newton's laws of motion.

An exploratory activity should precede each chapter in the course. Whether or not each student does the activity, or you do it as a group demonstration, will depend on equipment and time considerations. Chapter 2 Concept Development Practice Pages should be passed out to your class toward the end of your lecture or on the day following your lecture. Having your students do Practice Pages in class, either alone or with another student, enables mutual feedback.

In the following suggested lectures, keep in mind the old Chinese proverb, **"I hear and I forget; I see and I remember; I do and I understand."** So when discussing a falling object, let your students see you actually drop something — if only an eraser or a piece of chalk. When discussing speed, toss a baseball from hand to hand and secretly substitute a tennis ball when winding up, then pitch it to the class. Try to show your students something while you talk. Also, consider this: Your students will remember what they themselves are prompted to talk about, more than what you talk about. Be sure to use the "check-your-neighbor routine" discussed in the introduction. Get the students "talking physics." Very importantly, get them "doing physics" via exploratory activities, Practice Pages, and lab experiments. These ancillaries should be a strong part of the course.

I suggest that you have your students use the lab manual. Do Activity 2, *The Physics 500*, before your lecture, and follow your lecture and discussion of chapter material with either Activity 3, *The Domino Effect*, or Experiment 4, *Merrily We Roll Along*. Activity 5, *Conceptual Graphing*, utilizes an ultrasonic range finder like those found in automatic focus cameras, and it is the most powerful introduction to graphing that I have ever seen. If you have a computer, Activity 6 is the place to use the *Race Track* program.

Suggested Lecture

Starting with Motion: If you start with this chapter, you might acknowledge that a sensible way to begin the course is with simple concepts and then gradually build to more complicated concepts as the course progresses. However, you're *not* going to do it that way. Instead, you're going to begin with a description of motion that will be more quantitative than later material — serious stuff. This idea that the course does not become progressively more difficult should lessen the anxiety of students who already read Chapter 2 and found it intimidating. Although kinematics can be a difficult field to cover, your plow setting will not go too deep. Continue on a light note with the question: What means of motion has done more to change the way cities are built than any other? [The elevator!]

Neglecting Air Resistance: Explain the importance of motion as it pertains to the other areas of physics. State that such things as air resistance, buoyancy, spin, and the shape of a moving object are important considerations in the study of motion, and that beneath them are some very simple relationships which might otherwise be masked. These relationships are what Chapter 2 and your lecture are about. Add that completely neglecting the effects of air resistance on heavy and compact (dense) objects traveling at moderate speeds not only exposes these simple *relationships*, but it is a reasonable assumption. That is, one would notice no difference between the rate of fall of a heavy rock dropped from the classroom ceiling to the floor below when it fell through either air or a complete vacuum. For a feather and heavy objects moving at high speeds, air resistance does become important and will be treated in Chapter 4.

DEMONSTRATION [2-1]: Here's a simple and nice one. First, drop a sheet of paper and watch it flutter to the floor. Next, crumple the paper and watch it fall faster to the floor. Drop the two side by side. The effect of air resistance is obvious. Drop a sheet of paper and a book side by side. Of course the book falls faster due to its greater weight compared to air resistance. (Interestingly, the air resistance is greater for the faster-falling book, because it falls faster and plows more air out of its path than the piece of paper. This is an idea you'll return to in Chapter 4.) Now place the paper against the lower surface of the raised, horizontally-held book. When you drop them, nobody is surprised to see them fall together. The book pushed the paper with it. Now repeat with the paper on *top* of the book. Ask for the students' prediction and have them discuss it with their neighbor. Now surprise most of your class by showing them that the paper falls as fast as the book! (You may choose to tease your class and *not* show what happens; ask them to try it on their own after class.) The book will "plow through the air" leaving a path free of air-resistance for the paper to follow!

Speed: Define *speed* while you write its equation in longhand (speed = distance/time) on the board and give examples for measuring speed, such as automobile speedometers. Stress that the slash in your notation is a division sign that means "per", as in so many kilometers per so many hours. Next walk across the floor with 1-meter strides, one per second. Explain that you covered a distance of 1 meter per second, hence your speed is 1 m/s.

> CHECK QUESTIONS: If you ride a bike a distance of 5 m in 1 s, what is your speed? For 10 m in 2 s? For 100 m in 20 s? [Each answer is 5 m/s.] (Very important point: You'll improve your instruction appreciably if you allow some thinking time, say three seconds, after you ask a question. Not doing so is the folly of too many instructors.)

Velocity: Define velocity (speed with direction) in a similar manner. Tell your students that the distinction between speed and velocity is not critical in this chapter, because only cases along a straight-line path are being discussed. The distinction between speed and velocity will be more important when vectors are treated in Chapter 6. Point out the more important distinction between average speed or velocity and between instantaneous speed or velocity. It is the average speed for a particular trip that is usually calculated. The speedometer of a car reads the speed at any instant. These distinctions need not be belabored. They will make more sense when the need to discuss them arises in specific cases. The concepts that are important to distinguish are *velocity* and *acceleration*.

> CHECK QUESTIONS: If an airplane travels 500 km due north in 1 h, what is its velocity? 250 km due north in 1/2 h? 125 km north in 1/4 h? [Each answer is the same, 500 km/h north.]

Acceleration: On the board write the equation for *acceleration* (acceleration = **change** in velocity/time), and emphasize the word *change*. Cite how it is that the change in motion one feels while sitting in a vehicle when acceleration takes place. One lurches in the vehicle when it undergoes changes in speed or direction. State that there are three controls in an automobile that make the auto accelerate. Ask for them. (accelerator, brakes, and steering wheel) State that one undergoes acceleration when executing a curve, even though the speed may not change. The direction changes, so we see that the definition of velocity includes direction in order to make the definition of acceleration an all-encompassing one.

Numerical Examples: Give numerical examples of acceleration in units of kilometers per hour per second to establish the idea of acceleration. Be sure that your students are working on the examples with you. For example, ask them to find the acceleration of a car that goes from rest to 100 km/hr in 10 seconds. It is important not to use examples involving seconds again until the students taste success with easier examples that use kilometers per hour per second. Have them check their work with their neighbors as you go along. Only after your students get the hang of it should you introduce examples involving meters per second per second in order to develop a sense for the units m/s^2.

> CHECK QUESTIONS: What is the acceleration of a car that goes from 0 to 100 km/h in 10 s? [(10 km/h)/s] What is the acceleration of a mechanical part that moves from 0 to 10 m/s in a time of 1 s? [10 (m/s)/s, or 10 m/s^2]

10 m/s^2: Although the acceleration of free fall at the earth's surface is about 9.8 m/s^2, we round this off to 10 m/s^2 in order to establish more easily the relationship of velocity and distance. At the conclusion of your discussion, you can move to the more accurate 9.8 m/s^2 value, in accordance with the following chapters and lab work.

Free Fall: Drop an object from your outstretched hand and ask if it accelerated as it fell. Ask how much it accelerated, and then ask what it means to say it accelerated 10 m/s^2. Encourage neighbor-discussion. In specific words; In every second of fall, the object picks up 10 m/s more speed than it had the second before. This is made clearer if you suppose the falling object is equipped with a speedometer, as per the Practice Page for this chapter. Even without a knowledge of physics, most people know instinctively that the speed of a falling object increases with time. (That's why one wouldn't hesitate to catch a baseball dropped from a height of 1 meter, but one would be quite reluctant to catch the same baseball if it were dropped from an airplane at high altitude.) Presumably your students now understand that if a freely-falling object was equipped with a speedometer, the speed reading would increase by 10 m/s for each successive second of fall.

> CHECK QUESTION: If an object is dropped from a rest position at the top of a cliff, how *fast* will it be going after falling 1 second? [9.8 m/s] (After asking the check question, you might say, "Write the answer on your notepaper." Then add, "Look at your neighbor's paper. If your neighbor doesn't

have the right answer, reach over and help — talk about it." With some discretion say, "If your neighbor isn't cooperative, sit somewhere else next time!" Peer pressure, when properly used, can be an effective inducement toward an active rather than passive class.)

You may distinguish between the acceleration of free fall and the deceleration the ball undergoes when it strikes the floor. Deceleration upon impact is considerably more than 10 m/s², because the velocity change occurs in a much shorter time. Deceleration is simply negative acceleration.

Table 2-2: After explaining the answer to the above Check Question and when class discussion dies down, repeat the process and ask for the speed at the end of 2 seconds and 10 seconds. This leads you into stating the relation $v = gt$, which by now you can express in shorthand notation.

Summary: This should be enough information for one class period. After any questions, discussion, and examples, state that you are going to pose a different question in the next class period by asking not how *fast*, but how *far*. Ask your class to consider before the next class how far a freely-falling object falls in 1 second.

NEXT-TIME QUESTION: Consider leaving your class with the "Bikes and Bee" question which highlights the relationship $d = vt$ and is shown here. Display the question via an overhead transparency, or post it in an appropriate area for viewing. A glass case outside the classroom where other students can catch the flavor of conceptual physics is ideal!

Shown here are reduced versions of 8-1/2 x 11 sheets of paper that you can photocopy from the *Next-Time Questions Book*. See pages 5 and 6 in this book for more information about these ancillaries. At least one Next-Time Question is available for every chapter in the text, and each has a separate answer sheet. Next-Time Questions for other chapters are not shown in this guide.

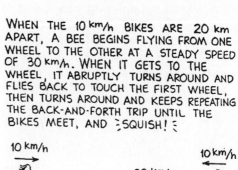

WHEN THE 10 km/h BIKES ARE 20 km APART, A BEE BEGINS FLYING FROM ONE WHEEL TO THE OTHER AT A STEADY SPEED OF 30 km/h. WHEN IT GETS TO THE WHEEL, IT ABRUPTLY TURNS AROUND AND FLIES BACK TO TOUCH THE FIRST WHEEL, THEN TURNS AROUND AND KEEPS REPEATING THE BACK-AND-FORTH TRIP UNTIL THE BIKES MEET, AND ⸘SQUISH!⸘

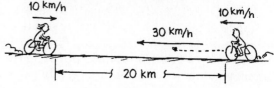

QUESTION

HOW MANY KILOMETERS DID THE BEE TRAVEL IN ITS TOTAL BACK-AND-FORTH TRIPS?

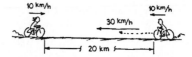

SOLUTION

LET THE EQUATION FOR DISTANCE BE A GUIDE TO THINKING:

$$d = \bar{v}\,t$$

WE KNOW $\bar{v} = 30$ km/h, AND WE MUST FIND THE TIME t. WE CONSIDER THE SAME TIME FOR THE BIKES AND SEE IT TAKES 1 HOUR FOR THEM TO MEET, SINCE EACH TRAVELS 10 km AT A SPEED OF 10 km/h. SO,

$$d = \bar{v}\,t = 30\ \text{km/h} \times 1\ \text{h} = 30\ \text{km}$$

THE BEE TRAVELED A TOTAL OF 30 km.

"Bikes and Bee" Solution: State the usefulness of the equation $d = vt$ in solving the "Bikes and the Bee" problem. Note that it is simpler when time is considered. Whether or not one thinks about time should not be a matter of cleverness or good insight, but rather a matter of letting the equation dictate the variables to be considered.

16

The velocity, v, is given but the time, t, is not. The formula instructs one to consider time. The key is that the same amount of time is required for both the bee's trip and the bike trip. (We neglect the short time of the bee's turnarounds.) Equations are important in guiding our thinking.

Review: If you're beginning a new class period at this point, review the essential points of the last meeting; the difference between average and instantaneous speed and velocity, the definition of acceleration, and the rule for finding how fast a freely-falling object falls in a given time, $v = gt$. While reviewing, ask how *fast* an object is falling 1 second after release from rest, after 2 seconds, after 10 seconds. Now pose the very different question, "How *far* does an object fall in 1 second?"

How Far: State that if the speed of fall is a steady 10 m/s for the entire second, how far it falls is entirely different than in the free-fall situation. Ask for a written response and then ask if the students could explain to their neighbors *why* the distance is only 5 m rather than 10 m. [In free fall, the object does not maintain a speed of 10 m/s throughout the second of fall. Its average speed during this time is only 5 m/s, where $v_{avg} = (v_f + v_i)/2$.] Point out that the relation $d = vt$ holds only when v is the *average* speed or velocity.

Table 2-3: Discuss the check question and answer on page 20 of the text, then call attention to Table 2-3 on page 20. Be sure your students see the difference between this table (which is about *distance* of free fall) and Table 2-2 (which is about *speed* of free fall). You may or may not deem it profitable to derive $d = 1/2\ gt^2$ on the board, as is done in the footnote on text page 20. (I usually do so, and I say that the derivation is a sidelight to the course — something that would be the crux of a follow-up physics course. In any event, the derivation is not something that I expect of them, but it is done to show that $d = 1/2\ gt^2$ is a statement of reason and that it doesn't just pop up from nowhere.)

CHECK QUESTION: How far will a freely-falling object that is released from rest fall in 2 seconds? In 10 seconds? (When your class is comfortable with this, ask how far it falls in 1/2 second.) [20 m, 500 m, 5/4 m]

To avoid information overload, I restrict all numerical examples of free fall to cases that begin at rest. Why? Because it's simpler that way. (I prefer that my students understand simple physics rather than be confused about not-so-

simple physics!) If you wish to go further, consider the initial speed to be greater than zero.

CHECK QUESTION: Consider a rifle fired straight downward from a high-altitude balloon. If the muzzle velocity is 100 m/s and air resistance can be neglected, what is the *acceleration* of the bullet after one second? [If most of your class say that it's g, you're on!]

What I do *not* do is to ask for the time of fall for a freely-falling object, given the distance. Why? Unless the distance given is the familiar 5 meters, algebraic manipulation is needed. If one of my teaching objectives was to teach algebra, this would be a nice place to do it. However, I don't have time to present this stumbling block and then teach how to overcome it. I'd rather put my energy *and theirs* into straight physics!

Nonzero Initial Speed:

CHECK QUESTION: Consider a rifle fired straight downward from a high-altitude balloon. If the muzzle velocity is 100 m/s and air resistance can be neglected, what is the velocity of the bullet after one second? [110 m/s]

The three activities at the end of the chapter, text pages 23 and 24, about measuring speed and reaction time are interesting and worthwhile classroom activities. Encourage students to do these either in or out of class.

Again, resist spending too much time on the material in Chapter 2. Don't get bogged down here. If most of your students have made the distinction between velocity and acceleration, move on to Newton's laws of motion and the tasty physics that follows.

More Think-and-Explain Questions

1. What is the acceleration of a vehicle that travels at a steady speed of 100 km/hr for 10 seconds? Explain your answer.
   Answer: The acceleration is zero, because no change in motion occurs. (It cannot be overemphasized that many mistakes are made in not carefully reading the question that is being asked!)

2. Suppose you stood at the edge of a cliff, as in text Figure 2-6, and threw a ball straight up with a certain speed. Suppose that you also threw another ball straight down with the same speed. Neglecting air resistance, which ball would have the greater speed in striking the ground below?

Answer: Although the ball thrown upward will be airborne for a longer time, both balls will strike the ground below with the same speed. When the upward-moving ball returns to its starting point, it will have the same downward speed as the ball that is thrown down.

3. If you drop an object in the absence of air resistance, its acceleration toward the ground is 9.8 m/s². If you throw it down instead, will its acceleration immediately after the throw be greater, the same, or less than 9.8 m/s²? Explain.

Answer: Although its instantaneous speed is greater at any time by the amount of its initial speed, the acceleration it experiences (a pickup of 10 m/s each second) is the same. (Speed and acceleration are entirely different concepts!)

4. Suppose you are watering a garden and holding the hose vertically so that the water stream rises 5 meters above the nozzle before falling back. What is the speed of the water as it leaves the nozzle?

Answer: 10 m/s; The water will fall back a distance of 5 m, which according to Table 2-2 takes 1 second. Movement up and movement down are symmetrical (text Figure 2-6), so 1 second is required for the water to go from the hose to the 5-m elevation. Consequently, the average speed of the water is 5 m/s. This means its initial speed is twice the average, or 10 m/s.

Computational Problems

1. How *far* will a freely-falling object released from rest fall in 1 second? In 2 seconds? In 10 seconds? In 1/2 second? In 1/10 second?

Answers: 5 m, 20 m, 500 m, 1.25 m, 0.05 m; by substitution into $d = 1/2\ gt^2 = 5t^2$: No answer is complete without units.

2. What will be the speed of a car that accelerates at 2 m/s² for 10 seconds from a position of rest?

Answer: 20 m/s; speed = acceleration x time = (2 m/s²)(10 s) = 20 m/s.

3. A ball rolls from rest down an inclined plane tilted at 30° to the horizontal, accelerates at 5 m/s² and reaches the bottom in 2 seconds. How long is the inclined plane?

Answer: 10 m; $d = 1/2\ at^2 = 1/2\ (5\ \text{m/s}^2)(2s)^2 = 10$ m.

4. If you throw a ball straight upward at a speed of 10 m/s, how long will it take for the ball to reach zero speed? How long will it take to return to its starting point? How fast will the ball be going when it returns to its starting point?

Answers: 1s; 2s; 10 m/s: Since it loses speed at a rate of 10 m/s each second, it takes 1 second for the ball to reach zero speed. The time the ball takes going up will be equal to the time it takes coming down. The speed the ball loses going up will be equal to the speed it gains coming down.

5. What is the speed of an apple that falls from a tree to the ground 5 meters below?

Answer: 10 m/s: To fall 5 meters takes 1 second, then $v = gt = (10\ \text{m/s}^2)(1\ s) = 10$ m/s or 9.8 m/s, since $g = 9.8$ m/s².

6. A dragster going at 15 m/s north increases its velocity to 25 m/s north in 4 seconds. What is its acceleration during this time interval?

Answer: 2.5 m/s²: Acceleration = (change in velocity)/(time change) = (25 - 15 m/s)/4s = 2.5 m/s².

7. A car going at 30 m/s undergoes an acceleration of 2 m/s² for 4 seconds. What is its final speed? How far did the car travel while it was accelerating?

Answers: Final speed is the initial speed v_0 plus that gained by accelerating; that is, $v = v_0 + at = 30\ \text{m/s} + (2\ \text{m/s}^2)(4\ s) = (30 + 8)$m/s = 38 m/s. Distance traveled is the distance the car would have gone in 4 s without accelerating, plus the additional distance ($1/2\ at^2$) it travels by accelerating, that is, $d = v_0t + 1/2\ at^2 = (30\ \text{m/s})(4\ s) + 1/2\ (2\ \text{m/s}^2)(4s)^2 = 120\ \text{m} + 16\ \text{m} = 136$ m.

8. We drive for 1 hour at 20 km/h, then we drive for 1 hour at 30 km/h. What is our average speed?

Answer: 25 km/h: average speed = distance traveled/time = (20 km + 30 km)/2h = 50 km/2h = 25 km/h.

9. We drive a distance of 1 kilometer at 20 km/h, then we drive an additional distance of 1 kilometer at 30 km/h. What is our average speed?

Answer: 24 km/h: Note that the time taken to travel the first kilometer is 1/20 hour and the time to travel the second kilometer is 1/30 hour. Average speed = distance traveled/time = (1 km + 1 km)/(1/20 h + 1/30 h) = 2 km/(3/60 + 2/60)h = 2 km/(5/60 h) = 24 km/h. Be careful that you don't simply average numbers when finding average speeds. The key is being sure to consider the time involved!

To use this planning guide work from left to right and top to bottom.

Chapter 3 Planning Guide
• *The bulleted items are key: Be sure to do them!*

Topic	Exploration	Concept Development	Application
Inertia	Act 7 (< 1 period)	• Text 3.1–3.6/Lecture Demos 3-1 to 3-6 Con Dev Pract Pg 3-1	Nx-Time Q 3-1 Act 8 (1 period) Nx-Time Q 3-2

Video: *Newton's First Law — Inertia*
Evaluation: Chapter 3 Test

See the Chapter Notes for alternative ways to use all these resources.

Objectives

After studying Chapter 3, students will be able to:
• Describe Aristotle's concepts of natural motion and violent motion.
• Describe Copernicus' idea about the earth's motion.
• Describe Galileo's contribution to the science of motion.
• Define inertia.
• State Newton's first law of motion.
• Distinguish among mass, volume, and weight.
• Distinguish between the kilogram and the newton as units of measurement.
• Explain how an object not connected to the ground can keep up with the moving earth.

Possible Misconceptions to Correct

• Constant motion requires a force.
• Even if no force acts on it, a moving object will eventually stop.
• Inertia is a force.
• Weight and mass are two names for the same thing.
• Mass and volume are two names for the same thing.

Demonstration Equipment

• [3-1] Coat hanger and clay blobs (pictured on next page)
• [3-2] Wooden block stapled to piece of cloth
• [3-3] Tablecloth with no hem and set of cheap dishes
• [3-4] Massive ball with hooks for strings above and below
• [3-5] Hammer and massive wooden block
• [3-6] Hammer and same wooden block

Introduction

A historical perspective is used to introduce the concept of inertia in this chapter. The historical figures considered are Aristotle, Copernicus, Galileo, and Newton. If you're a science history buff, you should consider amplifying the small amount of history in this text.

The videotape, *Inertia*, goes with this chapter. The basic ideas of the chapter are covered, along with some demonstrations that are not done so easily in class — such as banging on large blocks of wood and iron and demonstrating the inertia of a massive anvil placed on your stomach and then hit with a sledgehammer. You may show this tape as your lecture of the chapter, since it covers the material well enough to allow devoting more of your time and effort to discussion and lab preparation.

Do the Practice Page for this chapter after you have lectured about the chapter material.

Suggested Lecture

The Old Idea that Constant Motion Needs a Force: Begin by pointing to an object in the room and stating that if it started moving, one would reasonably look for a cause for its motion. We would say that a force of some kind, a push or a pull, was responsible. It doesn't seem that things move of themselves. Only living things do that. A cannonball remains at rest inside the cannon until a force is applied. The force of expanding gases drives the ball out of the barrel when the cannon is fired, but what keeps the ball moving when the expanding gases no longer act on it? This question leads to a discussion of inertia.

Aristotle and Galileo: Contrast the way Aristotle and Galileo looked at motion — Aristotle's classification of natural and violent motion and Galileo's idea of inertia, that objects once set in motion *do* continue moving without added effort. This is easy to understand when air resistance and other factors are not part of the situation, which occurs in outer space. It's easier to visualize the unchanging motion of a ball thrown in outer space than one thrown at the earth's surface. The ball may curve in outer space because of gravity, no air resistance to affects its motion.

Law of Inertia: Do the following demonstration before stating the law of inertia .

DEMONSTRATION [3-1]: Shape a wire coat hanger into an "m" as shown. Two globs of clay are stuck to each end. Balance it on your head with one glob in front of your face. State that you wish to view the other glob and ask how you can do so without touching the apparatus. Then simply turn around and there it is! It's like turning a bowl of soup only to find the soup stays put. Inertia in action!

State the law of inertia for objects in the rest case and then do the following demonstrations:

DEMONSTRATION [3-2]: Place a wooden block on a piece of cloth and ask what will happen to the block if you yank quickly on the cloth (the old table-cloth-and-dishes trick). But, for humor have the cloth stapled to the block beforehand. Then you're illustrating Newton's "zeroeth" law — Be skeptical!

DEMONSTRATION [3-3]: Do the classic table-cloth demonstration complete with dishes. It is important to pull slightly downward when you whip the table cloth from beneath the dishes. This insures that the cloth moves horizontally. Even the slightest upward component produces disaster. Even though your students may know what to expect, it is nevertheless well worth your while to do this. Seeing it live, and by you, is a delight!

Mass vs Volume: Distinguish also between mass and volume. Cite an automobile battery and a fluffy king-size pillow as examples. Mass is not volume, as Figure 3-8 in the text illustrates.

Mass vs Weight: Distinguish between mass and weight. Mass is more fundamental, meaning that it has to do only with matter and not with any external gravity. Weight is the force due to the earth's gravity on an object and depends on both gravity and matter. In a region where there is no gravity, there is no weight. However, the objects there still have mass. A useful way to impart the distinction between the two is to place two objects of about equal mass in the hands of a student. Ask the student to judge which is heaviest. If the student responds by shaking the objects back and forth with one in each hand, point out that the student is unconsciously comparing their *inertias* and is making use of the intuitive notion that mass and weight are directly proportional to each other. In the same region of the earth's gravitational field, twice the mass also has twice the weight. But this does not mean that mass is weight. (Twice as much sugar has twice the sweetening power, but this does not mean that sugar is sweetening power. Sugar *has* sweetening power, just as mass *has* weight in a gravitational field.)

DEMONSTRATION [3-4]: Perform the activity described in Think and Explain question 5 on page 36 of the text and also shown in the sketch. Ask for predictions. Then show that the bottom string breaks when it is jerked and that the top string breaks when the bottom string is gradually pulled harder. Ask for the explanation and use the "neighbor check" routine. [The answer is that the slow pull demonstrates the role of weight, since there is tension in the top string even before you pulled on it. This tension is increased as you pull down, and it is greater than the tension in the bottom string. By how much? By the weight of the ball. The quick jerk, on the other hand, demonstrates inertia, since the ball resists the sudden downward acceleration imposed by the lower string.]

DEMONSTRATION [3-5]: Place a massive block on your hand and strike the block with a hammer blow. You are not hurt because of the inertia of the block. Or do as I do in the videotape. Place an anvil on your stomach and invite a skillful person to whap it with a sledgehammer. In any event, be sure to show the relationship of this demo to the previous demonstration of the suspended ball and strings.

Emphasize the motion part of Newton's first law: Moving things continue at constant velocity in the absence of a force.

DEMONSTRATION [3-6]: Show how you tighten the head of a hammer (Figure B of Think and Explain question 6 on page 36 of the text) by banging the handle, not the head, against a solid surface, as shown in the sketch. Why? Because the massive head remains in motion and nestles tighter on the handle.

Shorter at Night: Relate the idea of tightening the hammer head to the bones of the human spine. As a result of jostling throughout the day, we are a bit shorter at night. The bones of the spine settle closer together, enough to be noticeable! Ask your students to find a place in their homes that is just out of reach of their up-stretched hands before they retire for the night. Tell them to try to reach the spot when they arise the next morning. They can't help but appreciate physics when they find that they can touch the place that was out of reach the night before! (Astronauts returning from orbit are more than an inch taller for the same reason — the absence of a force compressing an otherwise upright back. Short people trying to qualify for a minimum height have been known to stay in bed for days before having their height measured.)

Motion Without Force: Ask what the motion would be of a ball tossed in empty space and away from gravity and other forces? [Straight-line path at constant speed.] Ask what two words say the same thing as "straight-line path at constant speed?" [Constant velocity] Consider a rock swinging in a circle at the end of a string. What is the path of the rock when the string breaks? [A straight line, except for the influence of gravity. Note that before the string breaks, the rock swings at constant *speed*, not constant *velocity*. It is accelerating nonetheless because it is changing the direction part of velocity. We'll return to this idea in Chapter 9.]

Moving Earth: Stand next to a wall on the west side of your classroom. State that relative to the sun, the wall and the whole room is traveling at about 30 km/s. Aristotle's followers would say that you also move at 30 km/s, because you are standing on the moving floor. Now jump up. Ask why the wall didn't slam into you. After all, you can say that when you are no longer "attached" to the moving floor, there is no force making you travel along with the moving floor. Let your class discuss this before you point out that this is simply the bird-and-worm example discussed and answered in Section 3.6 in the text.

More Think-and-Explain Questions

1. Your empty hand is not harmed if it bangs lightly against a wall, but it is harmed if your are carrying a heavy load. Why is this so?

   Answer: A heavy load has a lot of mass and once in motion has a tendency to remain in motion. More force is required to stop a massive load, so your unfortunate hand is squashed in the process.

2. Does a person diet to lose *mass* or to lose *weight?*

    Answer: A person diets to lose mass. One loses weight whenever gravity is reduced, like on the surface of the moon. Even with less weight on the moon, an obese person is still obese.

3. Can the force of gravity on a 1-kilogram mass ever be more than on a 2-kilogram mass? Defend your answer.

    Answer: Yes: For example, the force of gravity on a 1-kg mass on the surface of the earth is considerably more than on a 2-kg mass on the surface of the moon (actually three times as much).

4. A car at a junk yard is compressed until its volume is less than 1 cubic meter. Has its mass changed? Has its weight changed? Explain.

    Answers: Neither its mass nor its weight changes when the car is compressed, because the same quantity of matter is present — only the volume is less.

5. If you jump up in bus that is moving at constant velocity, will you land farther back? Explain.

    Answer: No, you will land as you would if the bus were at rest. The explanation is that you are moving with the bus before, during, and after your jump. In accordance with Newton's first law, a body in motion remains in motion unless acted upon by a force. There is no horizontal force on you during your jump, so you simply continue moving with the bus. (If the bus accelerates while you are in the air, you land in a different place.)

6. If the first law is really a description of nature, why do moving things in our environment eventually slow down?

    Answer: Moving objects slow down because of forces, usually friction.

Computational Problems

1. An average apple weighs about 1 newton. What is its mass?

    Answer: 1/10 kg: Since 1 kg weighs about 10 newtons, we find 1 kg/10 N = x/1N; where x = (1kg)(1N)/10 N = 1/10 kg. At sea level on the earth's surface, 1 kg more precisely weighs 9.8 N. In the next chapter we will see that Weight = mg.

2. Find your weight on a bathroom scale and compute your mass in kilograms. If the scale reads pounds, compute your weight in newtons.

    Answer: Each kg weighs about 2.2 lb, or 10 N, so divide your weight in pounds by 2.2 and you have your mass of kilograms. Multiply the number of kilograms by 10 (or 9.8) and you have your weight in newtons. (This simple, but very good, activity helps to distinguish between mass and weight).

4 Newton's Second Law of Motion – Force and Acceleration

To use this planning guide work from left to right and top to bottom.

Chapter 4 Planning Guide
• *The bulleted items are key: Be sure to do them!*

Topic	Exploration	Concept Development	Application
Force and Acceleration	Act 9 (>1 period)	• Text 4.1–4.3/Lecture Demo 4.1, 4.2 Exp 10 (>1 period) Exp 11 (>1 period)	Nx-Time Q 4-1
Statics and Friction		• Text 4.4–4.5/Lecture Demo 4-3 Con Dev Pract Pg 4-1	Nx-Time Q 4-2
Pressure		Text 4.6 Lecture	Nx-Time Q 4-3
Falling		• Text 4.7–4.8/Lecture Con Dev Pract Pg 4-2	Nx-Time Q 4-4 Exp 12 (>1 period) Nx-Time Q 4-5

Video: *Newton's Second Law – Force and Acceleration*
Evaluation: Chapter 4 Test

See the Chapter Notes for alternative ways to use all these resources.

Objectives

After studying Chapter 4, students will be able to:
• Define net force.
• State the relationship between acceleration and net force.
• State the relationship between acceleration and mass.
• Distinguish between the concepts of *directly proportional* and *inversely proportional*.
• State Newton's second law of motion.
• Describe the effect of friction on a stationary object and on a moving object.
• Distinguish between force and pressure.
• Apply Newton's second law to explain why the acceleration of an object in free fall does not depend on the mass of the object.
• Describe what happens to the acceleration and the velocity of a falling object when there is air resistance.

Possible Misconceptions to Correct

• If an object has zero acceleration, it must be at rest.
• Pressure and force are the same.
• Heavy objects always fall faster than light objects do.
• Objects have no weight in a vacuum.

Demonstration Equipment

• [4-1] Cart pull (same apparatus of the lab experiments *Constant Force* and *Changing Mass*)
• [4-2] Spool and string

Introduction

Acceleration and inertia, introduced in Chapters 2 and 3, are developed further and are related to force in this chapter. Falling objects, as introduced in Chapter 2, are treated in more detail in this chapter.

Avoid placing too much emphasis on unit analysis. Don't get bogged down with the unit relationships $N = kg\ m/s^2$ and $m/s^2 = N/kg$. Belaboring these can be counterproductive, and they are best emphasized later in a student's study of physics — after concepts are understood and problem solving becomes more important. Your effort here is better placed on building a savvy of Newton's second law with interesting examples.

Magnitudes can be presented on the chalkboard by exaggerating the sizes of mathematical symbols rather than plugging in numerical quantities (see text pages 46 and 48). Rather than writing "change in" for equations on the board, introduce the symbol Δ, if you have not already done so (see the footnote on text page 37).

For your information, the terminal speed of a skydiver is about 60 m/s (215 km/h, or 134 mph), for a baseball it is about 42 m/s (150 km/h, or 95 mph), and for a Ping-Pong ball it is about 9 m/s (32 km/h or 21 mi/h). Drop Ping-Pong balls from small to large elevations and observe the heights to which they bounce upon impact with the floor. Beyond a certain elevation the balls bounce no higher, indicating they have reached their terminal speed!

Friction is treated briefly in this chapter, with an extended discussion in the Chapter 8 lab experiment, *Slip Stick*.

Precede your lecture with Lab Manual exercises. Use *Getting Pushy*, and follow up with *Get Up and Go* and *Keep on Truckin'*. These are a two-part sequence. *Impact Speed* is the most advanced experiment in the Lab Manual and is intended for your very best students. If you are into computers, this special topic can be pursued further with the *Laserpoint* computer program *Impact Speed*, which introduces the idea of area under the curve and sets the stage for integral calculus. Do not use *Impact Speed*, either the lab experiment or computer program, with students who are finding physics difficult. It will overwhelm them. Your whiz kids, on the other hand, will eat it up.

Suggested Lecture

Review Acceleration: Begin by reviewing examples and the definition of acceleration. Ask how one produces the acceleration of an object. On the board, write the idea that acceleration is imposed by an impressed force, $a \sim F$, and give examples of doubling and tripling the impressed force and the subsequent doubling and tripling of acceleration.

Net Force: Introduce the idea of net force by placing an object on your table and pushing it. State that neglecting friction, if you push it to the right with 10 N, the net force is 10 N. Ask what the net force would be if a student simultaneously pushed it to the left with 10 N. [0] With 4 N? [6 N to the right] State that the block would accelerate no differently if pushed with the two forces that produce a 6-N net force or a single applied 6-N force. The two are equivalent.

DEMONSTRATION [4-1]: Compare the accelerations of a large mass and a small mass when impressed upon by equal forces. Bring inertia into the discussion to complete Newton's second law. You might do this with the apparatus for the lab experiments *Constant Force and Changing Mass* and *Constant Mass and Changing Force*.

DEMONSTRATION [4-2]: If you're not going to assign the spool and string problem (Activity 3 on text page 52) as homework, demonstrate it here. Show that the acceleration is in the direction of the net force. The spool will roll toward you (providing your pull is horizontal) whether the string is wrapped over the top or under the bottom of the spindle. You can show this also on your chalkboard by using the sequence of sketches shown.

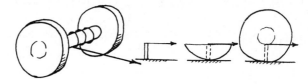

Statics: Call attention to the fact that the force of gravity acts on a book on the table, but the book is not accelerating. Ask why. To make credible the idea that the table is exerting an upward support force on the book, ask if an ant between the book and the table would be squashed on both its top and bottom sides. Or, ask if a compressed spring between the book and table would be squeezed from both the top and bottom. The table exerts upward support, called the *normal force*. As the name implies, the normal force is perpendicular to the supporting surface — always. When you weigh yourself on a weighing scale, your normal force is what you read. The reading differs if you try weighing yourself on an inclined surface. (For the record, the reading is your weight x *sin* Θ, where Θ is the incline from the horizontal). If you stand on a scale on a level surface, the normal force is your weight. Ask what the reading would be if you stood evenly distributed on two scales? Relate this to the idea that the net force on an object is zero whenever the acceleration of the object is zero.

CHECK QUESTION: How does the tension force in each of the little girl's arms in text Figure 4-7 compare to her weight? [Half the tension is on each arm.]

When two different conclusions are presented by two students for a given supposed situation, remind them that there are at least four ways to resolve the argument; (1) Impose the power of a higher authority, (2) Win by superior logic, (3) Take a vote, (4) Conduct an experiment and observe the result. In science the last choice trumps the others.

DEMONSTRATION [4-3]: Suspend a mass by a pulley that is supported by two strands of string as shown in the sketch. Show that the scale reading is half the weight of the mass. Ask what the reading would be if the mass were suspended by three strands? Ten? [1/3, 1/10]

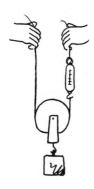

Sign-Painter Skit: On the board, draw the sketch shown of two painters on a painting rig suspended by two cables. Step 1: If both painters have the same weight and each stands next to a cable, the tension forces in the cables will be equal. Ask what the tension in each cable would be in this case. [The weight of one man plus half the weight of the rig. Why? Like the girl in Figure 4-7, the total support force upward must equal the total weight downward.] Step 2: Suppose one painter walks toward the other, as shown in the second sketch, which you duplicate on your chalkboard or overhead projector. Will the tension in the left cable increase? [Yes, because the left cable is sharing a greater portion of the total weight.] Will the tension in the right cable decrease? [Yes, because the left cable is doing most of the supporting.] Grand question: Will the

tension in the left cable increase exactly as much as the tension in the right cable decreases? [Yes, the increase in the tension in one cable exactly matches the decrease in the other. This complies with Newton's first and second laws — with the first law, the system at rest means no net force acts on it, and with the second law of motion, $a = F/m$. Since there is no acceleration of the supported system of rig and painters, the net force acting on the system must be zero. That means the upward support forces balance the downward forces of

gravity. Thus, a decrease in the tension of one cable is met with a corresponding increase in the tension of the other to produce a net force on the system of zero.] (This example is dear to my heart. When I was a sign painter and before I had any training in physics, sign painter Burl Grey posed this question to me. He didn't know the answer, nor did I. That was because neither he nor I had a model for analyzing the problem. We didn't know about Newton's laws, and therefore we didn't think in terms of zero acceleration and the corresponding zero net force. How different one's thinking is when one has, or does not have, a model to guide it. If my friend and I were into psuedoscience, we might have been more concerned with the notion of how each cable "knows" about the condition of the other. This is the approach that intrigues many people with a nonscientific view of the world.)

Nonaccelerated Motion: Pull a block across your lecture table with a spring balance so that everyone can see how much force you are applying. Pull the block so that it slides at constant velocity. Ask for the acceleration? [Zero, since there is no change in its velocity once it is moving steadily.] Ask what the applied force is. [The force showing on the scale.] Ask for the value of the net force acting on the block. [Zero, as evidenced by its nonchanging velocity.] Ask how the net force can be zero. [It is zero because of the friction between the block and the surface of the table.] Ask how much friction force acts on the block. [Equal to the scale reading but oppositely directed.] You are demonstrating the idea of a zero net force on a moving system. Stress the idea that zero acceleration applies both to the state of rest and the state of constant velocity. In both cases, there is no change in the state of motion. This is because in both cases the forces that act on the object balance to zero. Zero net force means zero acceleration. Zero acceleration is evidence of zero net force.

CHECK QUESTION: (See question 2 on page 43) Suppose the captain of a high-flying airplane announces over the public address system that the plane is flying at a constant speed of 900 km/h and the thrust of the engines is a constant 80 000 newtons. What is the acceleration of the airplane? [It's zero, because velocity is constant.] What is the combined force of air resistance that acts all over the plane's outside surface? [It is 80 000 N to produce a zero net force. If resistance were less, the plane would speed up; if it were more, the plane would slow down.]

Draw a free-body diagram on the board to illustrate the foregoing. This sets the stage for the treatment of vectors in Chapter 6.

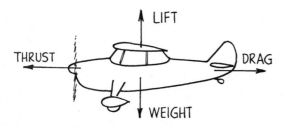

Pressure: Return to the block that you pulled across the table with the spring balance. Drag it with the wide side down, then with the narrow side down. Show that the force required to pull it uniformly across the table is independent of which surface of the block is against the table. Lead into the idea that in both cases the weight of the block against the table is the same, but the *distribution of weight* is different. Define pressure. Conclude this part of your lecture with examples; sharp points versus blunt surfaces, being stuck with a boxing glove versus a bare fist, taking too many courses and too many social activities at one time.

The lab experiment "Inflation" scheduled for Chapter 12 in the lab manual can be moved here if you wish to go further with pressure. It involves computing the weight of a car by measuring the area of contact of its tires on the pavement. This

idea with regard to a person's weight is discussed in this chapter on page 44.

(This is a good place to break.)

Falling Objects: Drop a book and a sheet of paper to show the different falling rates. Then crumple the piece of paper and show that it and the book accelerate about equally when dropped. State that air resistance for the low speeds involved does not reveal itself in your demonstration. State also that you are not going to talk about the effects of air resistance until you have first investigated the physics that occurs in the absence of air resistance. Too-early a preoccupation with air resistance can hide some very basic physics. (Here is where your self discipline may be challenged by an overanxious student who asks questions about air drag anyway. Defer such questions until the simplest case is understood by your class.) State that Galileo, who is reputed to be the first to publicly show equal accelerations for unequal masses, could not adequately explain why. He lacked the model offered by Newton, namely Newton's second law.

LECTURE SKIT: Parallel the comic strip on text page 47 by pretending you are experimenting with two bright youngsters. Hold a kilogram weight and a piece of chalk above your head and ask the class which will hit the ground first if you release them at the same time. [Your class should answer, "the same."] Ask your class to pretend you are asking this question to a youngster who has not been exposed to this idea before. Then carefully articulate a good argument for the heavy weight falling first (as Charles Schultz might ask of his little characters in "Peanuts").[The kilogram weight is pulled more by gravity than the chalk.] Go through the motions of asking the same question of another child, who responds with a good argument for the chalk hitting first. [The kilogram has more inertia than the chalk and will take more time to get moving than the chalk.] Summarize the first child's argument by writing the equation, $a \sim F$, on the board. Summarize the second child's argument with the equation $a \sim 1/m$. State that the beauty of science is that such speculations can be determined by experiment. Drop the weight and the chalk to show that no matter how sound each child's argument seemed to be, the results do not support either. Then bring both arguments together with $a = F/m$, Newton's second law. Relate this to the case of the falling cannonball and stone in text Figure 4-14.

Falling and Air Resistance: Thus far you have avoided the effects of air resistance in your discussion (if you've been successful in fending off this consideration until after you have discussed the simpler case first — that of free fall). Newton's second law will still be the model for investigating falling in the presence of air resistance. The only difference will be that the net force is not the weight, but the weight minus air resistance (see the note at the top of text page 49).

The "Principle of Exaggeration:" In discussing the effects of air resistance on falling objects, it is useful to exaggerate the circumstances so that the effects are more clearly visualized. For example, in comparing the falls of a heavy and a light skydiver, ask your students to substitute the falling of a feather for the light person and the falling of a heavy rock for the heavy person. It is easy to see that the air resistance plays a more significant role for the falling feather than it does for the falling rock. To some extent, this is similar to the falls of the two skydivers.

Terminal Speed: Discuss text Figures 4-15 and 4-16 (the skydivers and the flying squirrel). State that the air resistance that a falling body encounters depends on two things, (1) how big it is, or the size of the path of plowed air, like the difference between the air resistance that acts on a flying squirrel and a regular squirrel and (2) how fast it's going. Ask if anyone has ever put their head out the window of a fast-moving car and noticed that the faster the car moves, the greater the "wind" is. Add that if the speed is great enough, the wind will clean the dandruff from one's hair! Air resistance depends on both size and speed. As an object falls faster and faster, there may be a point where the force of air resistance is equal to the weight of the falling body.

> CHECK QUESTIONS: When the force of air resistance on a falling object is equal to the object's weight, what will be the net force on the object? [Zero] What will the acceleration be then? [Zero] Does this mean the falling object comes to an abrupt halt; that is, that zero acceleration means zero velocity? [No, zero acceleration does not mean zero velocity, but zero CHANGE in velocity.]

Non-Free Fall: Thus far you have discussed two cases of falling; falling with no air resistance with acceleration *g*, and falling at a terminal velocity with acceleration zero. Analyzing accelerations between *g* and zero for falling objects requires more thought. Use your class to gauge how far you wish to explore the subject matter on the last page of the chapter (text page 50). You can either share the anno at the top of page 49 with your class and direct discussion to this equation, or describe how the net force on a falling object will decrease as air resistance builds up to decrease the accelerating effect of the object's weight. If a skydiver steps from a high-altitude balloon (at rest to reduce the variables involved), she will initially accelerate downward at *g*. When she reaches terminal speed, her acceleration will be zero. What happens between the time she first jumps and reaches terminal speed? [Her acceleration must have diminished as she gained speed, and therefore air resistance. As long as motion is in a straight line, the terms terminal speed and terminal velocity are interchangeable.]

> CHECK QUESTIONS: A skydiver jumps from a high-altitude balloon. As she falls faster and faster through the air, does air resistance increase, decrease, or remain the same? [Increase] Does the net force on her increase, decrease, or remain the same? [Decrease, because the net force on her is her weight minus air resistance. As air resistance increases, net force decreases.] As she falls faster and faster, does her acceleration increase, decrease, or remain the same? [Acceleration decreases, because net force decreases. When she falls fast enough her acceleration will reach zero and she will have reached terminal velocity.]

More Think-and-Explain Questions

1. Consider an object that doesn't accelerate when acted on by a force. What inference can be made?

   Answer: Another force, such as friction, must be counteracting the first force.

2. When a car moves along the highway at constant velocity, the net force on it is zero. Why then does the engine continue to burn fuel?

   Answer: The engine burns fuel to supply the force to overcome air resistance and other friction forces, which combine to produce the zero net force.

3. What is the net force on an apple that weighs 1 N when it is held at rest above your head? What is the net force on it when it is released?

   Answers: Net force is zero when it is at rest; net force equals 1 N, its weight, when it is in free fall.

4. Why is a soft couch more comfortable to sit on?

   Answer: Compared to a hard surface, there is more area of contact when sitting on a soft couch. The greater area results in less pressure and more comfort for you.

5. How does the weight of a falling object compare to air resistance just before it reaches terminal velocity? How does it compare after terminal velocity is reached?

   Answers: Just before reaching terminal velocity, weight is slightly greater than air resistance, so it is still accelerating. After reaching terminal velocity, air resistance has built up to equal weight, so acceleration becomes zero (terminates).

6. In the absence of air resistance if a ball is thrown vertically upward with a certain initial speed, will it have the same speed on returning to its original level. Why?

   Answer: It will have the same speed because the "deceleration" while ascending is the same as the acceleration while falling. The speed it loses going upward, is the same as it gains coming down.

7. In the preceding question if air resistance is a factor, will the speed of the ball reaching the ground be greater than, less than, or the same as the speed without air resistance?

   Answer: The speed of the ball will be less, because air resistance opposes its motion. This is easy to see with the "principle of exaggeration." Suppose the ball is a feather cast upward into the air. Its speed is quickly diminished because of air resistance, and it flutters slowly to the ground with much less speed than it began with.

8. In the preceeding question, would you expect the time ascending to be greater, less, or the same as the time descending in the presence of air resistance?

   Answer: The ball will ascend in less time than it falls. Like the preceding question and answer, this is best seen by the principle of exaggeration. The time for the feather to fall from maximum altitude is clearly longer than the time it took to attain that altitude. The same is true for the not-so-obvious case of the ball.

Computational Problems

1. If a 1-N net force accelerates a 1-kg mass at 1 m/s^2, what is the acceleration of a net force of 2 N on 2 kg?

   Answer: The same; $a = F/m = 2$ N/2 kg = 1 N/1 kg = 1 m/s^2.

2. What is the acceleration of a 747 jumbo jet (mass is 30 000 kg) in takeoff when the thrust for each of its four engines is 30 000 N?

   Answer: $a = F/m = (4 \times 30\ 000$ N)/30 000 kg = 4 m/s^2.

3. A certain force applied to a 2-kg mass accelerates it at 3 m/s^2. How much acceleration will the same force produce on a 4-kg mass?

   Answer: The acceleration will be half as much, 1.5 m/s^2, because the same force acts on twice the mass. $F = ma = (2$ kg)(3 m/s^2) = 6 kg m/s^2 = 6 N. So $a = F/m = 6$ N/4 kg = 1.5 N/kg = 1.5 m/s^2. Note that the units check; kg is mass, m/s^2 is acceleration, and mass times acceleration equals force, N, so (kg m/s^2) = N.

4. An occupant of a car has a chance of surviving a crash if the deceleration is not more than 30 gs. Calculate the force on a 70-kg person at this rate.

   Answer: $F = ma = (70$ kg)(30 x 9.8 m/s^2) = 20 580 N. This is about 4620 pounds.

5. What will be the acceleration of a skydiver when air resistance is half the weight of the skydiver?

   Answer: It is $g/2$, because the net force $(W - R)$ is half that of free fall; $a = F/m = (W - R)/m = (W - W/2)/m = W/m - W/2m$. Since $W/m = g$, we see that $a = g - g/2 = g/2$.

6. A 10-kg mass on a horizontal friction-free air track is accelerated by a string attached to another 10-kg mass hanging vertically from a pulley. What is the weight in newtons of the hanging 10-kg mass? What is the acceleration of the system of both masses in m/s2?

   Answers: The weight of the 10-kg mass; weight = $mg = (10$ kg)(9.8 m/s^2) = 98 N. The acceleration of the system(recall that both masses are accelerating); $a = F/m = 98$ N/20 kg = 4.9 m/s^2. This is one-half g, which is to be expected because the weight of one mass accelerates twice as much mass.

7. In the preceding problem, suppose the masses are 1 kg and 100 kg. Compare the accelerations when they are interchanged, that is for the case where the l-kg mass dangles over the pulley and the 100-kg mass dangles over the pulley. What does this indicate about the maximum acceleration of such a system of masses?

   Answer: For the case of the smaller mass dangling over the pulley, the accelerating force is its weight = $mg = (1$ kg)(9.8 m/s^2) = 9.8 N. The acceleration of the system is this force divided by the (1 + 100)-kg mass of the system; $a = F/m = 9.8$ N/101 kg = 0.1 m/s^2. When the masses are interchanged, the accelerating force is the weight of the larger mass = mg = (100 kg)(9.8 m/s^2) = 980 N; the acceleration of the system $a = F/m = 980$ N/101 kg = 9.8 m/s^2 (rounded-off value). From this we see that the maximum acceleration of such a system is that of free fall, g. No combination of masses can produce an acceleration outside the range 0 to 9.8 m/s^2.

Newton's Third Law of Motion – Action and Reaction

To use this planning guide work from left to right and top to bottom.

Chapter 5 Planning Guide

• The bulleted items are key: Be sure to do them!

Topic	Exploration	Concept Development	Application
Newton's Third Law	Act 13 (1 period)	• Text 5.1–5.7/Lecture Con Dev Pract Pg 5-1 Exp 14 (>1 period) Exp 15 (1 period)	Nx-Time Q 5-1 Nx-Time Q 5-2 Nx-Time Q 5-3

Video: *Newton's Third Law–Action and Reaction*
Evaluation: Chapter 5 Test

See the Chapter Notes for alternative ways to use all these resources.

Objectives

After studying Chapter 5, students will be able to:
- Define force in terms of interaction.
- Explain why at least two objects are involved whenever a force acts.
- State Newton's third law of motion.
- Given an action force, identify the reaction force.
- Explain why the accelerations caused by an action force and by a reaction force do not have to be equal.
- Explain why an action force is not cancelled by the reaction force.

Possible Misconceptions to Correct

- Pushes and pulls are applied only by living things.
- Things such as high-speed bullets contain force.
- Reaction forces occur slightly after the action force is applied.
- Action and reaction forces are equal and opposite only under certain conditions.

Demonstration Equipment

None required.

Introduction

Many people think of a "reaction" as the response to a given force. For example, if someone gives you a shove, you respond by shoving back. When the earth pulls on the moon, the moon responds by pulling back with the same magnitude of force. However, that's not what Newton's third law is about. Strictly speaking, bodies don't respond or "react" to pushes and pulls. Pushes and pulls are simultaneous interactions between the bodies concerned.

Newton's third law leads nicely into momentum conservation. All the examples that illustrate action and reaction also serve to illustrate the conservation of momentum. For example, we say that the kick of a fired gun is the reaction to the force on the bullet, but we could just as well say that the momentum of the kicking gun matches the momentum of the bullet. If one started with momentum conservation, one could lead nicely into Newton's third law. Either topic may be considered fundamental. For the time being, if you wish to go directly from this chapter to Chapter 7 *Momentum* and skip Chapter 6 *Vectors* there is good reason to do so. Chapter 6 separates Newton's third law and momentum, because the vector nature of momentum is briefly treated in Chapter 7. Consequently, there is better reason to follow the Chapters 5, 6, and 7 sequence of the text. The choice is yours.

Do both the exploratory activity *Tension* and the experiment *Tug-of-War* from the Lab Manual before you lecture on this chapter. *Balloon Rockets* can be done either with this chapter or just before the following chapter on momentum.

Suggested Lecture

Begin by reaching out to your class and stating, "I can't touch you without you touching me in return — I can't nudge this chair without the chair nudging me — I can't exert a force on a body without that body exerting a force on me. In all these cases of contact, there is a twoness — contact requires two bodies." Then state Newton's third law and support it with examples.

Examples of Newton's Third Law: Extend your hand and show your class that you can bend back your fingers a only very little distance. Show that if you push with your other hand and thereby apply a force to them, they will bend appreciably more. Walk over to the wall and as you push against the wall, show that the inanimate wall does the same to you. State that everybody will acknowledge that you are pushing on the wall, but only physics types realize that the wall is simultaneously pushing on you also — as evidenced by your bent fingers. State that this is just one of the fundamental physical laws that completely escape the notice of most people, even though its staring them right in the face. We look without seeing. State that much of physics is pointing out the obvious that is all around us.

When walking you interact with the floor, that is you push back on the floor and the floor pushes forward on you. When swimming you interact with the water, that is you push backward on the water and the water pushes forward on you. A balloon pushes escaping air backward and the escaping air in turn pushes the balloon forward, just as a jet plane does in flight. A car pushes backward on the road, and the road pushes forward on the car. It is the road that pushes a car along!

CHECK QUESTIONS: Suppose two bowling balls A and B are separated and connected by a stretched spring. Is A pulling on B, is B pulling on A, or are both pulling on each other? [They are part of the same interaction and they pull on each other.]

Identifying Action and Reaction: Call attention to the examples in Figure 5-4 on page 56

of the text and relate these to the interaction rule.

Object A exerts a force on object B.
Object B exerts a force on object A.

Discuss the action and reaction pairs of forces when a bullet is fired from a rifle. Use the exaggerated symbol technique on text page 57 to show how equal forces produce unequal accelerations when different masses are involved. I cite the imagined case of a person who is found guilty of rational thinking in a totaliarian society and is about to be shot by a firing squad. He is given one last wish. Because his punishment should be harsher than being struck by a tiny bullet, his last wish is that the mass of the bullet be much greater than the gun from which it is fired — and that his antagonist pull the trigger!

Tug-of-War: Ask your students to pretend a tug-of-war is taking place. One team pulls their end of the rope with 1000 newtons of force and the other team pulls their end of the rope similarily with 1000 newtons. Ask what the tension will be in the rope, if it could be shown by a spring balance in the middle. Give the options in multiple choice form; (1) 2000 N, (2) 1000 N, or (3) 0 N. After several seconds of class discussion and without giving the answer, dismiss the answer of zero newtons. Do this by reminding your class that if a person were to take the place of the rope and two teams pulled equally on each arm, the tension would be appreciably more than zero! The rope really is being stretched. Ask the class to pretend one team gets tired, so you call a halt to the tug-of-war. You tie their end of the rope to the wall, out of view of the other team. Pretend you drape a curtain between the groups. When the game resumes, will the pulling team be able to tell that the other team has been replaced by a wall? They may attribute a great "steadiness" to the other team, but when they pull as before, everything will seem the same. Ask now what the scale reading will be? It should be clear that it will read 1000 N and not 2000 N, as many will presume. The rope pulls just as a team does. When you pull on the wall, the wall really does pull back on you!

Points about a tug-of-war: The winning team is not the team that pulls hardest on the rope, but rather the team that pushes hardest against the ground. To win in a tug-of-war, you can (1) maintain your stance until your opponent tires and lessens his push against the ground, (2) push harder against the ground than your opponent does and maintain rope tension, or (3) push harder against the ground than your opponent does and increase rope tension.

CHECK QUESTION: Apply Newton's third law to the tug-of-war. If the action is you pulling on the rope, is the reaction force the ground pushing back on you or your opponent pulling back on the rope? [Neither, reaction is the rope pulling back on you — A on B; B on A!]

Why Action and Reaction Forces Don't Cancel: Call attention to Figure 5-11 on text page 59, which shows the apple and orange pulling on each other. Action and reaction are like apples and oranges in that they act on different objects. You can't cancel a force on an orange with a force on an apple. This is the essence of the horse-and-cart problem.

Horse-Cart Problem: Discuss action and reaction in terms of the horse and cart situation described on page 61 and the comic strip, "Horse Sense," on text page 60. Point out that the action force on the cart is not canceled by the reaction because the reaction acts on the horse. This is the apple and orange example of Figure 5-11. You can't cancel a force on the cart by a force on something else.

Difficulties that occur with action-and-reaction situations usually stem from failing to clearly identify the *system* in question. (Read ahead to page 40 in this manual for more about defining systems.) Basically, if you want to know the effect of the force or forces on something, call that something your system. Define your system by a real or imaginary dotted line around that something. Restrict your attention to only the external forces that originate outside the dotted line and act on the system, and not to the forces that the system may exert on things beyond the dotted line. We distinguish between the forces ON a system and the forces BY the system on other objects. Consider the horse-cart problem. If the system is the cart, then the only horizontal force to act on the cart is the pull of the horse — period! Consequently, there is a net force on the cart and acceleration occurs. If the system is the horse, draw a dotted line around the horse. Two horizontal external forces act on this system, the reaction by the cart and the reaction by the ground (friction) due to the horse's push against the ground. If the ground push is greater than the cart's pull, the horse accelerates (just as much as the joined cart!) If the system is both the horse and the cart, then only one horizontal external force acts on this system; the same push of the ground. Divide this push by the mass of the horse and cart, and you have the acceleration of both (the same as before).

Paper Punch: Drop a sheet of paper and then punch it in midair. State that the heavyweight boxing champion of the world couldn't hit the paper with a force of 50 pounds. No way! That's because the paper is not capable of "hitting back" with the same amount of force. A 50-pound interaction between his fist and the paper isn't possible.

More Think-and-Explain Questions

1. Consider the two forces acting on the person who is standing still; namely the downward pull of gravity and the upward support of the floor. Are these forces equal and opposite? Do they comprise an action-reaction pair? Why or why not?

   Answers: The forces are equal and opposite because they are the only forces acting on the person, who obviously is not accelerating. They cancel to produce the zero net force that nonacceleration requires. Note that they do not comprise an action-reaction pair however, because they are not the parts of a single interaction. There are two interactions in question here; the interaction between (1) the person and earth and (2) the floor and the person. Interaction (1) is weight; the earth pulls down on the person (action) and the person pulls up on the earth (reaction). (Note that the action and reaction are co-parts of a single interaction that acts on different bodies). Interaction (2) is upward support by the floor on the person (action) and downward push by the person on the floor (reaction). Again, action and reaction act on different bodies in a single interaction, but the equal and opposite forces cited in the question act on the same body.

2. You exert 200 N on your refrigerator and push it across the kitchen floor at constant velocity. What friction force acts between the refrigerator and the floor? Is the friction force equal and opposite to your 200-N push? Does the friction force make up the reaction force to your push?

   Answers: Constant velocity means zero acceleration which means zero net force. Therefore, the friction force must be equal and opposite to your 200-N push, but it does not make up the reaction to your push. The reaction to your push against the refrigerator is the refrigerator pushing back on you. You and the fridge make up one interaction, and the floor and the fridge make up another. Even though the magnitudes of all forces are the same, there are two interactions with two sets of action-reaction pairs of forces.

6 Vectors

To use this planning guide work from left to right and top to bottom.

Chapter 6 Planning Guide
• The bulleted items are key: Be sure to do them!

Topic	Exploration	Concept Development	Application
Vectors	• Act 16 (1 period)	• Text 6.1–6.4/Lecture Exp 17 (>1 period)	
Equilibrium		• Text 6.5–6.7/Lecture Demos 6-1, 6-2, 6-3 Con Dev Pract Pg 6-1	
Projectiles		• Text 6.8–6.9/Lecture Demos 6-4, 6-5 Con Dev Pract Pg 6-2, 6-3	• Exp 18 (>1 period) Nx-Time Q 6-1

Video: *None*
Evaluation: Chapter 6 Test

See the Chapter Notes for alternative ways to use all these resources.

Objectives

After studying Chapter 6, students will be able to:

- Distinguish between a vector quantity and a scalar quantity, and give examples of each.
- Draw vector diagrams of forces or velocities.
- Use the parallelogram method to find the resultant of two vectors that have different directions.
- Explain why a clothesline or wire, which can easily support an object when strung vertically, may break when strung horizontally and supporting the same object.
- Given a vector, resolve it into horizontal and vertical components.
- For an object on a slope, resolve its weight into a component that causes acceleration along the slope and a component that presses it against the slope.
- Compare the motion of an object which is rolled off a horizontal table to that of an object dropped from rest at the same time.
- Explain why a projectile moves equal distances horizontally in equal time intervals, when air resistance is negligible.
- For a projectile, describe the changes in the horizontal and vertical components of its velocity, when air resistance is negligible.

Possible Misconceptions to Correct

- 1 + 1 = 2, always.
- The tension in a string that supports a load is generally equal to the weight of the load.
- The curved motion of a projectile is very complicated.
- An object at rest will drop to the ground faster than the same object moving horizontally at high speed.
- At the top of its trajectory, the velocity of a projectile is always momentarily zero.

Demonstration Equipment

- [6-1] A pair of scales, string, and a mass of about 1 kg
- [6-2] A heavy chain that is several meters long
- [6-3] Sailcart and fan
- [6-4] Spring gun apparatus to simultaneously project a ball horizontally while dropping another from rest
- [6-5] Monkey-and-hunter apparatus

Introduction

A full treatment of vectors is often a formidable stumbling block for students. Vectors can be combined into a resultant or resolved into components by geometrical graphics. Using trigonometry is not necessary, but it is more efficient and precise. This chapter does not cover vectors entirely, it samples only some simple and interesting vector applications. This way your students should not see the material in this chapter as a stumbling block. Vectors are combined and resolved via the parallelogram rule, and the only trigonometry hinted at is the Pythagorean Theorem (see footnote on text page 71).

Vector analysis is a quite different and new mode of thinking for many students. Some may struggle before becoming comfortable with it. It is best to separate clearly the operations of geometric combination and resolution. Do this by spending one lecture period discussing only vector combination, and a different lecture period discussing vector resolution. A demonstration of a sailboat sailing into the wind is a very effective way to introduce vector resolution.

The interesting fact that projectiles launched at a particular angle have the same range if launched at the complementary angle is stated without proof on text page 81, and is shown in Figure 6-18. This fact is a consequence of the range formula, $R = (2v^2 sin\Theta cos\Theta)/g$, which is symmetrical for sine and cosine. Since the sine of an angle is the cosine of the complement of that angle, replacing the angle with its complement will give the same range. So the range is the same whether aiming at 0° or at 90° – 0°. Maximum range occurs at a projection angle of 45°, where sine and cosine are equal.

For your personal information and at the risk of contributing to "information overload" with your students, there is a simple rule for estimating how much less than 45° to project when the landing spot is below the launching elevation, like downhill. Simply subtract *half* the angle of the "incline" from 45°. For example, if the landing spot is 10° below the launch point, project at (45° - 5°) 40° for maximum horizontal displacement. If the landing spot is above the launching point, say uphill, then add half the angle of "incline" above the launch point to 45°. So if your projectile is to land uphill by 10°, launch it at 50°. This rule holds for maximum horizontal distance from launch point to landing point when air resistance is not important. This and other interesting tidbits about the projectile motion of baseballs, footballs, and frisbees can be found in the delightful book, *Sport Science*, by Peter J. Brancazio (Simon & Schuster, 1984).

Before you treat Section 6.4, have your class do the activity *24-Hour Towing Service*. Follow this up with a separate lecture begining with section 6.5. If you want to cover vectors extensively, continue to text Appendix D and demonstrate a sailboat sailing into the wind. This is a fascinating and powerful demonstration of vector resolution. Do this with a sailcart, preferably on an air track (as in the lab *Riding With the Wind*). A separate lecture period should cover projectiles, which are found in Section 6.8.

Suggested Lecture

Begin by calling attention to the fact that on a windy day one can run faster when running with the wind rather than against it. Similarly, a plane is often late, or early, in arriving at its destination due to wind conditions. Flying with the wind results in an increase in ground speed. Represent speed with an arrow. Since you are representing magnitude by the length of the arrow and the direction by the arrowhead, you are now talking about velocity — a vector quantity. Discuss text Figure 6-2 and variations of wind conditions that are only with or against the motion of the aircraft. Parallel this with a similar treatment of boats sailing with and against the stream.

> CHECK QUESTION: How fast would an airplane move over the ground if it has an airspeed of 100 km/h and is flying into the headwind of a strong gale of 100 km/h? [It would have a ground speed of zero, similar to birds flying into a strong wind and not going anywhere.]

Vectors at an Angle: Continue with the airplane and wind and consider a right-angle wind, as shown in Figure 6-5. Follow this up with a boat sailing across a stream, only consider 90 degree cases and invoke the Pythagorean Theorem — like 3-4-5 triangles for a start. After discussing the geometry of the square in Figure 6-6, ask the following question.

> CHECK QUESTION: How fast will a boat which normally travels 10 km/h in still water be moving with respect to land if it sails directly across a stream that flows at 10 km/h? [14.14 km/h.]

Non-Right Angle Vectors: Consider an airplane flying sideways to a wind that does not meet it at 90 degrees. The Pathagorean Theorem cannot be used for non-right angle situations — at least not directly. More analysis is needed. Introduce the technique of parallelogram construction, first for rectangles as in Figure 6-4, and then for vectors like those in the exercises on the same text page (70). Have your class construct parallelograms on the Chapter 6 Practice Pages. Check their work before going further.

(This is a good place to break.)

Begin with this demonstration.

DEMONSTRATION [6-1]: Set up a pair of scales that support a heavy weight as shown in Figure 6-8 on text page 72. A 1-kg mass is suitable. Show that as the supporting angle increases, the tension also increases, see Figure 6-9.

Figure 6-9 Explanation: Explain why the tension increases with the increasing angle by showing that the resultant of tensions in each strand must combine to a vector equal and opposite the weight vector. Carefully explain Figure 6-8.

DEMONSTRATION [6-2]: Have two students hold the ends of a heavy chain. Ask them to pull it horizontally in order to make it as straight as possible. Call attention to this and the clothesline of Figure 6-7. Next ask what happens if a bird comes along and sits in the middle (as you place a 1-kg hook mass on the middle of the chain!) What happens if another bird comes to join the first (as you add another 1-kg mass). Ask the students to keep the chain level. Now what happens if a flock of birds join the others (as you add additional masses). This works well! (Note that this parallels the lab *24-Hour Towing Service*.)

Relate this situation to the questions on text page 73 and ask why there is always a sag, however slight, in a horizontally-stretched rope or wire. Why? It's because the weight of the rope or wire must be matched by upward components of the tension along the rope or wire. The rope or wire must be directed slightly upward in order to provide the needed vertical component to offset the weight.

Vector Components: For a standard treatment, discuss Figures 6-10 and 6-12 and then direct students to the exercises of vector components in their Practice Pages book. For a fuller treatment, move to text Appendix D and discuss the section on sailboats.

DEMONSTRATION [6-3]: Give a demonstraton of a sailcart sailing with the wind, with a crosswind, and then angling into the wind. If you do this, spend the remainder of

the class period on sailboat discussion. Then follow this up with the lab experiment, *Sailboat Physics*.

If you don't get into sailboats, you can move into projectile motion, which is subject matter for a lecture of its own.

(This is a good place to break.)

Projectile Motion: Discuss the idea of the "downwardness" of gravity and how there is no "sidewaysness" to it. Consider a bowling ball rolling along a bowling alley. Gravity pulls it downward and completely perpendicular to the alley with no horizontal component of force, even if it rolls off the edge of the alley like a ball rolling off a tabletop. Pose the situation of the horizonally-held gun and the shooter who drops a bullet at the same time he pulls the trigger. Ask which bullet hits the ground first. [Neglecting the earth's curvature, both hit the ground at the same time.]

DEMONSTRATION [6-4]: Show the independence of horizontal and vertical motions with a spring-gun apparatus that will shoot a ball horizontally while at the same time dropping another ball that falls vertically. If you don't have such an apparatus, place a coin at the edge of a table and slide another coin across the tabletop to knock it off. The struck coin should fly across the room while the other coin more or less falls straight downward. Students will see that they hit the floor at the same time. Afterwards announce, "Gravity does not take a holiday on moving objects."

Independence of Horizontal and Vertical Motion: Ask for an explanation of how the photograph in Figure 6-14 was taken. [A strobe light was flashed in rapid bursts in a dark room, while the device dropped one ball and simultaneously projected the other horizontally. The photo was taken by a camera with its shutter held open during this time.] Compare the downward motions of each ball and note they are the same. Investigate the sideways motion of the projected ball and see that it moves equal horizontal distances in equal times.

CHECK QUESTION: How is the horizontal component of motion affected by the vertical component of motion? [It isn't! The horizontal and vertical components of motion are independent of each other.]

Point to some target at the far side of your clasroom and ask your class to imagine that you are going to project a rock to the target via a slingshot. Ask if you should aim directly at the target, above it, or below it. Easy stuff. Then ask your class to assume that it takes 1 second for the rock to reach the target. If you aim directly at the target, it will fall beneath and miss. How far beneath, if the floor wasn't in the way? Do a neighbor check on this question. When the class agrees it is 5 meters (or 4.9 m), ask how far above the target you should aim to hit the target. [The same 5 m or 4.9 m.] After another neighbor check you're ready to discuss Figure 6-15 on text page 79.

Upwardly-Moving Projectiles: Investigate Figure 6-15. Call attention to the vertical distances the projectiles have fallen and Table 2-3 on text page 20. (This is the same physics of free fall, only "stretched out horizontally.")

DEMONSTRATION [6-5]: Perform the famous "Monkey and Hunter Demonstration" if it is available. An alternative is to show the same sequence with a crossbow from the old PSSC film, *Projectile Motion*.

CHECK QUESTION: If the cannon were aimed downward instead of upward in Figure 6-15, how would the distances below the new "dashed line" compare? [The projectile displacements below the dashed line would be no different; 5 m at the end of the first second, 20 m at the end of the second second, and so on. (This is treated in the Practice Pages.)]

Investigate the relative vectors in Figure 6-16. Note that the horizontal component doesn't change, since no horizontal force acts. Note that the vertical component does change by going upward against gravity then downward with gravity. Note that the same is true of the steeper angle of Figure 6-17.

CHECK QUESTIONS: True or False? The velocity of projectile at its highest point is zero. [False, the vertical component of velocity is zero at the highest point, and not the velocity itself (unless it is projected straight upward).] What can be said of the velocity of the projectile at its highest point? [At its highest point and neglecting air resistance, the velocity of a projectile will be the same as its horizontal component of velocity is at any other point.]

Projectile Ranges: The equal ranges for projectiles launched at complementary angles is

quite interesting. Leave it at that for now. The explanation has to do with sine and cosine trig functions, so let that wait until a future physics course. Save your students' banks of grey matter for more fertile ground.

CHECK QUESTION: To direct water to flowers that are farthest away, at what angle should a water hose be held? [Ideally 45°; somewhat less if you're holding it high off the ground.]

Air Resistance: Acknowledge the large effect that air resistance has on the foregoing analysis, particularly for fast-moving objects such as bullets and cannonballs. A batted baseball, for example, travels only about 60 percent as far in air as it would in a vacuum. Its curved path is no longer a parabola, as Figure 6-20 indicates.

Satellite Motion: When the earth's curvature is taken into account and the launch speed is great enough, a projectile can become an earth satellite. Time permitting, you can end your lecture by acknowledging this fact. The physics is summed up in the comic strip, "Satellite Physics," found on page 192. Done properly, you can delightfully tease your class with the interesting physics that is yet to come!

Physics of Surfing: This feature on text page 83 can be further explained as follows. Hold a meter stick at an angle of about 30° or so above your table. Slowly lower the stick, maintaining its angle, so that it misses the table's edge. Ask your class to note the point of "contact" of the stick with the table — the point that moves horizontally across the surface as the stick is lowered. Ask for a comparison of the speed of the stick and the speed of the contact point. They should see that the point moves about twice as fast. Tip the stick to about 10° with the horizontal surface and repeat the procedure The point moves faster. A similar thing happens for the surfer who angles across the crest of a moving wave!

More Think-and-Explain Questions

1. Why does vertically-falling rain make slanted streaks on the side windows of a moving automobile? If the streaks make an angle of 45°, what does this tell you about the relative speeds of the car and the falling rain?

   Answer: The rain has a horizontal component of motion *relative to the moving car*, which when combined with the vertical motion of the rain, produces the slanted streaks. A 45° angle means that the speeds of the rain and car are the same.

2. A newspaper boy walks a series of 100-m city blocks. His route takes him 5 blocks east, then 3 blocks north, then 4 blocks west, and finally 2 blocks south. How far away is he from his starting point?

Answer: 141.4 m; His total east-west distance is 5 - 4 = 1 block east. His total north-south distance is 3 - 2 = 1 block north. This puts him one diagonal block northeast from his starting point, 141.4 m. We say his *displacement* is 141.4 m, quite different than his total walking distance of 1400 m.

Computational Problems

1. Harry Hacker is playing shuffleboard on an ocean liner that is travelling due north at 3 m/s on the blue Carribean. He makes a starboard shot at 4 m/s so that relative to the deck, the puck is moving east. What is the velocity of the puck relative to the stationary stars?

Answer: 5 m/s in a northeasterly direction; 37 degrees east of north, to be exact. The path of the puck makes the hypotenuse of a 3-4-5 right triangle.

2. John and Ted look from their 80-m high-rise balcony to a swimming pool below — not directly below, but rather 20 m from the bottom of their building. They wonder how fast they would have to jump horizontally to succeed in reaching the pool. What is the answer?

Answer: 5 m/s: At 5 m/s, they would reach the pool's closest edge. An 80-m drop requires 4 s, so they have 4 seconds to fall while they are moving horizontally. We see this from the equation for falling distance; $80 = 1/2\ gt^2$, $t = \sqrt{2(80)/g} = \sqrt{160/10} = 4$ s. In 4 s, they must cover a horizontal distance of 20 m to reach the edge of the pool. So their horizontal speed (horizontal distance/time) must be 20 m/4 s = 5 m/s. To land squarely inside the pool, somewhat more than 5 m/s is needed for success.

3. If the pool in the above question is 20 m long, estimate John's and Ted's upper limit of initial horizontal velocity for success?

Answer: 10 m/s: If the men jumped twice as fast, 10 m/s, they would go twice as far in the same time compared to the previous questions, and land 40 m from the building at the far edge of the pool. Ouch! Even a meter from the edge would put them in trouble; 2 meters may work, but 4 m would be more sensible. This puts the horizontal distance at 36 m. Then initial horizontal speed = 36 m/4 s = 9 m/s. The liklihood of overshooting the pool is small compared to whether they can muster the 5 m/s or more to reach the near side of the pool. Better to do this high jump in safer territory!

4. A cannonball shot with an initial velocity of 141 m/s at an angle of 45° follows a parabolic path and hits a balloon at the top of its trajectory. Neglecting air resistance, how fast is the cannonball going when it hits the balloon? What is its acceleration just before it hits the balloon?

Answer: 100 m/s: g: At the top of its trajectory, the vertical component of velocity is zero, leaving only the horizontal component. The horizontal component at the top, or anywhere along the path, is the same as the initial horizontal component, 100 m/s (the side of a square where the diagonal is 141). The acceleration at the top, or at any other point along the path, is that due to gravity, g.

7 Momentum

To use this planning guide work from left to right and top to bottom.

Chapter 7 Planning Guide
• *The bulleted items are key: Be sure to do them!*

Topic	Exploration	Concept Development	Application
Impulse /Momentum		• Text 7.1–7.2/Lecture Demo 7-1	
Bouncing	Act 19 (<1 period)	Text 7.3/Lecture Demo 7-2	
Conservation of Momentum		• Text 7.4/Lecture Con Dev Pract Pg 7-1	Exp 20 (1 period)
Collisions		Text 7.5–7.6/Lecture Demo 7-3	Nx-Time Q 7.1

Video: *Momentum*
Evaluation: Chapter 7 Test

See the Chapter Notes for alternative ways to use all these resources.

Objectives

After studying Chapter 7, students will be able to:
• Define momentum.
• Define impulse and relate it to momentum.
• Give examples of how both the size of the force and the length of the time interval affect the change in momentum.
• Explain why impulses are greater when an object bounces than when it simply comes to a complete stop.
• State the law of conservation of momentum.
• Distinguish between an elastic collision and an inelastic collision.
• Give an example of how the vector nature of momentum affects the law of conservation of momentum.

Possible Misconceptions to Correct

• Impulse equals momentum rather than a change in momentum.
• Momentum is conserved only when collisions are perfectly elastic.
• Impact and impulse are the same.

Demonstration Equipment

• [7-1] Sheet to hang and eggs to throw in it
• [7-2] Bouncing-dart apparatus, if your students don't do the activity *Bouncing Dart*.
• [7-3] Air track and carts of equal and unequal masses

Introduction

It is interesting to note that Newton expressed his second law in terms of momentum ($F=\Delta P/\Delta t$) rather than the familiar $F = ma$. (P is the usual symbol for momentum and is not used in this text until Chapter 16.) Many interactions that are explained by Newton's third law are explained by momentum conservation as well. Newton's laws flow nicely into momentum and its conservation. If it weren't for the need to treat vectors, this chapter would directly follow the chapters on Newton's laws. It is interesting to note that either Newton's third law or momentum conservation can be considered fundamental. That is, momentum conservation can be a consequence of Newton's third law, or equally, Newton's third law can be a consequence of momentum conservation.

We emphasize the impulse-momentum relationship with applications selected to catch student interest. In presenting your own application, the exaggerated symbol technique shown in text Figures 7-3, 7-4, and 7-5 is suggested.

Angular momentum is not treated until Chapter 11.

The popular swinging balls apparatus shown in the sketch provides an excellent illustration of momentum conservation. Students can easily see that when the balls on one side are lifted and then released so that they make contact with the others, the momentum of the balls is the same before and after the collision. The same number of balls emerges at the same speed on the other side, so momentum before collision is seen to be equal to the momentum after collision. However, the question is often raised, "Why cannot two balls be raised and allowed to swing into the array, and one ball emerge with twice the speed?" Be careful here. Momentum would indeed be conserved if this were the case, but the case with different numbers of balls emerging never happens. Why? [Energy would not be conserved.] For the two-balls-one-ball case, the KE after impact would be twice as much as the KE before impact. KE is proportional to the square of the speed, and the conservation of both momentum and KE cannot occur unless the numbers of balls for collision and ejection are the same. So if you include this demonstration, treat it in the next chapter instead of here.

Applications of momentum conservation often result in confusion if the idea of a *system* and the *isolation* of that system are not clear. The momentum of a system is conserved when no external forces act on the system. Defining and isolating that system is therefore important. Consider a cue ball that makes a head-on collision with an 8-ball at rest. If the system is taken to be only the 8-ball at rest, then we isolate it with a dotted border around it (sketch I). So long as no outside force acts on the 8-ball, there will be no impulse on it and no change in its momentum.

However, when the cue ball strikes it, there is an outside force and an impulse on the 8-ball. Its momentum changes as it speeds away with the speed of the incident cue ball. Or, take the system to be the cue ball (sketch II). Initially it has momentum mv. Then it strikes the 8-ball and its momentum undergoes a change. The reaction force by the 8-ball brings it to a halt. Now consider the system of both balls (sketch III). Before collision the momentum is that of the moving cue ball. When the balls strike no outside force acts, because the interaction is between the balls, which are both parts of the same system. So no impulse acts on the system and no change in momentum of the system occurs. In this case, momentum is conserved. Momentum is the same before and after the collision. Again the momentum of a system is conserved only when no external impulse is exerted on the system. Forces that are internal to a system do not change the momentum of the system.

I.

II.

III.

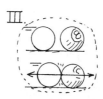

A system is isolated not only in space, but in time also. When we say that momentum is conserved when one pool ball strikes the other, we mean that momentum is conserved during the brief duration of interaction when outside forces can be neglected. After the interaction, friction quite soon brings both balls to a halt. So when we isolate a system for purposes of analysis, we isolate both in space and in time.

You may want to assign an "Egg Drop" experiment. Students design and construct a case to hold an egg that does not break when dropped from a three- or four-story building. The design cannot include means to increase air resistance, so all cases should strike the ground with about the same speed. By requiring the mass to be the same for every egg case, the impulse upon impact will be the same also. The force of impact, of course, should be minimized by maximizing the time of impact. This project is one that stirs considerable interest, both for your students and others who are not (yet?) taking your class. Here's an interesting sidelight. Loose heavy blankets hung in the windows of buildings in bullet-and-shrapnel-strewn areas are effective in slowing bullets and shrapnel.

Begin this chapter without a pre-lecture activity. The activity, *Bouncing Dart*, should be done after you discuss the impulse and momentum relationship and just before you treat *bouncing*. As a follow-up, have your class do the experiment *Tailgated by a Dart* in the lab manual.

Suggested Lecture

Momentum: Begin by stating that there is something different between a truck and a roller skate; they have different amounts of mass. Next state that there is also something different about a moving truck and a moving roller skate; each has a different *momentum*. Define and discuss momentum as moving mass; inertia in motion.

CHECK QUESTION: After stating that a truck has more mass than a roller skate, ask if the truck will always have more momentum than the roller skate? [No, if a truck is at rest it has no momentum. If a roller skate moves, it has momentum and thus more momentum than a truck at rest does.] Next ask if a truck and a roller skate could have the same nonzero momentum? How? [A truck moving very slowly and a roller skate moving very fast could have the same momentum. When this occurs, the product of mass times speed for both will be equal.]

Cite the case of huge oil-carrying supertankers that normally cut off their power when they are 25 or so kilometers from port. Because of their huge momentum (due mostly to their huge mass), about 25 kilometers with water resistance is required to bring them to a halt.

Impulse-Momentum: Write the impulse-momentum equation on the board. You may want to derive it on a remote part of the board (see footnote on text page 88). Make a distinction between impulse and *impact*. [Impact is a force.] Support the impulse-momentum relationship with examples found in the text. Other examples are (1) for increasing momentum — the long barrel of a cannon designed for long-range projection or the long strand of rubber for a long-range slingshot; (2) for decreasing momentum over a long time — pulling your hand backward when catching a fast

ball, jumping into a safety net as compared to concrete pavement, falling on the soft surface of a regular playing field as compared to the hard surface of Astroturf® recreational surface, falling on a wooden floor with give as compared to falling on street pavement; and (3) for decreasing momentum over a short time — any collision with a hard surface with little "give."

CHECK QUESTION: Why is falling on a wooden floor in a roller rink less dangerous than falling on the concrete pavement? [The wooden floor has more "give." Emphasize that this is the beginning of a fuller answer, one that is prompted if the question is reworded as follows.] Why is falling on a floor with more give less dangerous than falling on a floor with less give? [The floor with more give allows a greater time for the momentum of fall to become zero, and a greater time of impulse means less force.]

DEMONSTRATION [7-1]: Throw an egg in a sagging sheet as stated in Think and Explain Question 3 in the text.

Railroad Cars: The loose coupling between railroad cars provides a very good example of the impulse-momentum relationship. The loose slack in the coupling of railroad cars is evident when a locomotive either brings a long train into motion from rest, or when it brings a moving train to rest. In both cases, a cascade of clanks is heard as each car is engaged in turn. Why the loose coupling? Without it, a locomotive might simply sit still and spin its wheels. The friction force between the wheels and the track is simply inadequate to set the entire mass of the train in motion. There is, however, enough friction to set one car in motion, so the slack allows the locomotive to get one car going. When the coupling is tight, the next car is set in motion. When the coupling for two cars is tight, the third car is set in motion, and so on until the whole train is given momentum. So the slack allows the required impulse to be broken into a series of smaller impulses, or in effect, an extension of time so that the friction between the locomotive wheels and the track can do the job.

Karate: Confusion point: In boxing, for maximum impact one "follows through", whereas in karate it is commonly thought that one "pulls back." However, this is not so — a karate expert may pull back for the purpose of repositioning and maneuvering, but not for imparting a maximum blow. Maximum impact force occurs when the target is struck in such a way that the hand bounces back, not pulls back. This yields up to twice the impulse to the target (just as a ball bouncing off a rigid wall delivers nearly twice the impulse to the wall than it would if it stuck to the wall).

Bouncing: Discuss, as the text does, how Lester Pelton made a fortune by applying some simple physics to the old paddle wheels. Wind bouncing off a sail in a similar manner provides additional impulse to sailcraft. If you are not going to have your student do the activity *Bouncing Dart*, show it as a classroom demonstration.

(This is a good place to break.)

Conservation of Momentum: Distinguish between external forces and internal forces; for example, the difference between sitting inside a car and pushing on the dashboard and standing outside and pushing against the outside of the car. Only an external force will produce a change in the momentum of the car. No change in momentum can occur in the absence of an external net force. That is, when $F = 0$, $mv = 0$ also.

Isolating Systems: Discuss the idea of isolating a system when applying the conservation of momentum. We isolate a system in space by imagining a dotted boundary line around the perimeter of the system, and we isolate a system in time by considering only the duration of the interaction. Show that where momentum may be conserved for a particular system, it may not be conserved for part of the system (for example, the pool balls previously illustrated).

For another example, consider a dropped rock in free fall. If the system is taken to be the rock (sketch I), then momentum is not conserved as it falls because an external force acts on the system. (Its vector is seen to penetrate the dotted border of the system.) This external force, gravity, produces an impulse on the rock that changes its momentum. If the system is instead considered to be the rock plus the entire world (sketch II), then the interaction between the rock and the world is internal to the system. (There is no penetrating vector.) For this larger system, momentum is conserved. That is, the momentum of the world as it "races up" to meet the

falling rock is equal and opposite to the momentum of the rock as it drops to meet the world (at its center of mass). The momentum of any interaction is always conserved if you make your system big enough. As far as we can tell, the momentum of the universe is without change.

Collisions: When no external forces act on a system, no change in the momentum of that system occurs. This is best seen in collision. Distinguish between elastic (bouncy) and inelastic (sticky) collisions.

> DEMONSTRATION [7-2]: Show examples of momentum conservation in both elastic and inelastic collisions with carts on an air track. If you represent the collisions on the board, it will be sufficient to use the exaggerated symbol technique (big *m*, little *v*, and vice versa).

For the case of equal-mass carts in an inelastic collision (Figure 7-10 on text page 96), go over the equation at the bottom of text page 95 in detail. Write the similar equations for collisions you demonstrate on the air track so that students will relate the equations to visual examples. Have your students write the equations for the other examples you show. In writing the equations for head-on collisions, be careful to show the velocities in one direction as positive and the oppositely-directed velocities as negative. For example, the equation for Figure 7-9b is $[mv + m(-v)]_{before} = [m(-v) + mv]_{after}$. In both the before and after cases, the net momentum is zero.

Momentum Vectors: Illustrate the vector nature of momentum by discussing Figures 7-2 to 7-4 in the text. Resist making a big deal out of this section unless you have ample time on your hands and your class is anxious for you to set your academic plow deeper. I think it is enough for students to be exposed to the general idea here and then to move on.

More Think-and-Explain Questions

1. A fully-dressed person is at rest in the middle of a pond on perfectly smooth ice and must get to shore. How can this be accomplished?

   Answer: An article of clothing can be thrown in a direction opposite to the direction the person wishes to slide. The momentum given to the clothing will be offset by an equal and opposite momentum of the person. Another technique is to blow air like a jet. (When breathing in again, be careful to tilt the head downward or upward so that sliding is not slowed by inhaling!)

2. Your friend says that momentum conservation is violated when a ball rolls down a hill and gains speed. What do you say?

Answer: The ball gains momentum because of the impulse provided by the force of gravity. It is in the absence of an external force that momentum doesn't change. If the whole earth and the ball are taken together as a system, then the gravitational interaction between the ball and the earth is internal and no external force acts. The momentum of the ball is accompanied by an equal and opposite momentum of the earth and no net change in momentum results.

3. If a huge truck and a motorcycle have a head-on collision, which vehicle will experience the greater force of impact? The greater impulse? The greater change in momentum? The greater acceleration? And hence, the greater damage?

Answer: The magnitude of force, impulse, and change in momentum will be the same for each. The motorcycle undergoes the greater deceleration because its mass is less. The motorcyclist is in more trouble than the truck driver: "It's not the fall that hurts you; it's the sudden stop!" Note that the value of this question is in understanding the meanings of the terms *force*, *impulse*, *momentum*, and *acceleration*.

4. A 1000-kg car moving at 20 m/s slams into a stone wall and comes to a halt. Here are two questions to consider. (a) What impulse acts on the car? (b) What is the force of impact on the car? Which of these questions can be answered directly and which cannot? Explain.

Answer: Question (a) has a direct answer; 20 000 N (because impulse = momentum change). Question (b) cannot be answered without knowing the *time* of impulse, in which case the average force could be calculated. (In the next chapter we will see that knowing the *distance* traveled during impact would also allow solution of average force.)

Computational Problems

1. A car with a mass of 1100 kg moves at 24 m/s. What braking force is needed to bring the car to a halt in 20 s?

Answer: From $Ft = \Delta mv$, $F = \Delta mv/t = $ (1100 kg x 24 m/s)/20 s = 1320 N. Here the change in momentum is simply the momentum, 1100 x 24 kg m/s, the car has before braking occurs.

2. A 100-kg quarterback is traveling 5 m/s and is stopped in 1 s by a tackler. Calculate (a) the initial momentum of the quarterback, (b) the impulse imparted by the tackler, and (c) the average force exerted by the tackler.

Answers: (a) mv = (100 kg)(5 m/s) = 500 kg m/s. (b) Impulse = $Ft = \Delta mv$, so impulse = 500 N. (c) $F = \Delta mv/t$ = 500 Ns/1s = 500 N.

3. A jet engine gets its thrust by taking in air, heating and compressing it, and then ejecting it at a high speed. If a particular engine takes in 20 kg of air per second at 100 m/s and ejects it at 500 m/s, calculate the thrust of the engine.

Answer: Thrust (applied force) = 8000 N. From $Ft = \Delta mv$; $F = \Delta mv/t$ = (20 kg x 500 m/s) - (20 kg x 100 m/s) /1 s = 20 kg (500 - 100) m/s/1 s = 20 kg (400 m/s)/1 s = 8000 N.

4. A 40-kg projectile leaves a 2000-kg launcher with a speed of 400 m/s. What is the recoil speed of the launcher?

Answer: The launcher will have to have a recoil momentum that is equal and opposite to that of the projectile. The momentum of the projectile is 40 kg x 400 m/s = 16 000 kg m/s. For the launcher, 2000 kg x v = 16 000 kg m/s, where v = 16 000 kg m/s /2000 kg = 8 m/s.

5. A car with a mass of 700 kg travels at 20 m/s and collides with a stationary truck with a mass of 1400 kg. The two vehicles interlock as a result of the collision and slide along the icy road. What is the velocity of the car-truck system?

Answer: Momentum before is all in the car; 700 kg x 20 m/s = 14 000 kg m/s. The momentum after is the same, but consists of the mass of both vehicles, 2100 kg. So 2100 kg x v = 14 000 kg m/s, where v = 14 000 kg m/s/2100 kg = 6.6 m/s.

6. A 1-kg dart moving horizontally at 10 m/s impacts with and sticks to a wood block with a mass of 9 kg, and then slides across a friction-free level surface. What is the speed of the wood and the dart after they collide?

Answer: 1 m/s; 1 kg x 10 m/s + 0 = (9 + 1) kg x v; where v = 10/10 = 1 m/s.

7. A 10 000-kg vehicle moving at 10 m/s collides with a 2000-kg car moving at 30 m/s in the opposite direction. If they stick together after impact, how fast and in what direction will they be moving?

Answer: 3.33 m/s in the same direction as the 10 000-kg vehicle. Momentum before collision = (10 000 kg x 10 m/s) + (2000 kg x [-30 m/s]) = 100 000 kg m/s - 60 000 kg m/s = 40 000 kg m/s in the direction of the more massive vehicle. This is also the momentum after collision; momentum = (10 000 + 2000) kg m/s x v, the combined mass moving at velocity v. Solving we find v = 40 000/12 000 = 3.33 m/s.

8. In a fireworks display a 3-kg body moving at 4 km/h due north explodes into 3 equal pieces: A moves east at 4 km/h ; B moves 37° south of west at 5 km/h; and C moves due north at 15 km/h. After the explosion, what is the total momentum of all the pieces?

Answer: The same as before explosion, 3 x 4 = 12 units of momentum! If you wish to do this the long way, the momentum of piece A is cancelled by the east-west component of B, which is also 4 units. Note that B makes a 3-4-5 right triangle and the east-west Side 4 opposes A. This leaves a 3-unit, due-south component of momentum for B, which opposes the 15 units due north of A. So the total momentum is 15–3 = 12 units due north, exactly the momentum the system had to begin with!

8 Energy

To use this planning guide work from left to right and top to bottom.

Chapter 8 Planning Guide

• *The bulleted items are key: Be sure to do them!*

Topic	Exploration	Concept Development	Application
Work	Act 21 (1 period)	• Text 8.1/Lecture	Nx-Time Q 8-1
Power and Energy	Act 22 (>1 period)	• Text 8.2–8.5/Lecture	
Conversation of Energy	Exp 23 (1 period) Exp 28 (>1 period)	• Text 8.6/Lecture Demo 8-1, 8-2 Exp 24 (1 period) Exp 27 (1 period) Con Dev Pract Pg 8-1	Exp 26 (>1 period)
Machines		• Text 8.7–8.9/Lecture Demo 8-3 Exp 29 (1 period)	Exp 25 (1 period)

Video: *Energy*
Evaluation: Chapter 8 Test

See the Chapter Notes for alternative ways to use all these resources.

Objectives

After studying Chapter 8, students will be able to:
• Determine the amount of work done, given the force and the distance moved. Determine the amount of power required, given the work and the time.
• Define work in terms of energy.
• Distinguish among mechanical energy, potential energy, and kinetic energy.
• Give examples of situations in which (a) the gravitational potential energy changes and (b) it does not change even though something is moved.
• Describe how the kinetic energy of an object depends upon the speed of the object.
• State the law of conservation of energy.
• Describe the function of a lever.
• Give examples of situations in which the mechanical advantage of a machine is (a) greater than 1 and (b) less than 1.
• Explain why no machine can have an efficiency of 100%.

Possible Misconceptions to Correct

• Momentum and kinetic energy are much the same concept.
• Energy is conserved only under certain conditions.
• It is possible to get more energy out of a machine than is put into it.

Demonstration Equipment

• [8-1] Simple pendulum, say a small weight at the end of a 1-m length of string
• [8-2] Hand-cranked generator that lights a light bulb
• [8-3] Spring balance and model pulleys, as in text Figure 8-12

Introduction

This is an important chapter. The concept of energy is central to physics and is discussed in various forms throughout the text. The concept of mechanical energy is elusive, because it becomes evident only when it changes from one form to another or when there is motion. Also mechanical energy is relative, that is, it depends on the location we choose for our reference frame. A 1-newton apple held 1 meter above the floor has 1 joule of potential energy, but when held out the window 10 meters above the ground, the apple has 10 joules. The same apple held in your lap has zero kinetic energy, but if your lap is on the seat of a high-flying jet plane, the apple has many joules of kinetic energy relative to the ground below. Potential and kinetic energies are relative to a specified or an implied frame of reference.

According to my friend Dave Wall, a former patent office employee, the greatest shortcoming of would-be inventors is their lack of understanding of the law of energy conservation. The patent office has long been besieged with schemes that promise to circumvent energy conservation. This point is addressed in the question about fuel economy on text page 109, and it is well worth discussing. There are still charlatans who pretend to have invented energy-producing machines and who sell stock shares to people who are misinformed about the conservation of energy. Your students should not be among this group.

Levers and pulleys are treated briefly. If there wasn't so much other good physics to cover, you could spend several lectures on simple machines. I recommend you give it one lecture period at most. The last section of the chapter, *Energy for Life*, can be expanded in your lecture.

An interesting lecture presentation is found at the beginning of Chapter 4 in the first volume of *The Feynman Lectures on Physics* (Addison-Wesley, 1963). Feynman compares the idea of energy conservation with a child's misplaced blocks. The analogy is good resource material for class discussions.

Suggested Lecture

Fashion a long pendulum that extends to the ceiling and is two or three meters from a wall. Begin by standing on a chair against the wall with the extended heavy pendulum bob held at the tip of your nose, or against your teeth. Say nothing. Release the bob and let it swing out and then back to your face. Don't flinch. Then comment on your confidence in one of the most central of the physical laws — *the conservation of energy*. The study of energy begins with a related concept, *work*. (If you can't do this, describe the situation.)

Work: Define work and relate it to the lifting of a barbell, as shown in text Figure 8-1. When work is done on the barbell, two things happen; (1) a force is exerted on the barbell and (2) the barbell is moved by that force. If the barbell is simply held still, the weight lifter would get tired and feel like he is doing work. He may well be doing work on himself via tiny movements in his body tissues, but he is doing no work *on the barbell* unless the force he exerts *moves* the barbell. Distinguish between work done on an object and the work done by an object on something else. Also distinguish between work done by you, on you, and in you.

CHECK QUESTION: Work is done lifting a barbell. How much more work is done lifting a twice-as-heavy barbell the same distance? [Twice as much] How much more work is done lifting a twice-as-heavy barbell twice as far? [Four times the work done lifting the first barbell]

Potential Energy: Attach a spring balance to the pendulum bob at its equilibrium position. Show how a small force pulls the bob sideways from its equilibrium position. Make a comparison between this force and the force that would be necessary to lift the bob vertically (its weight). Show how as the bob is pulled farther up the arc, the force to move it increases. This is so because it is being pulled against gravity, which has no vector component along the pendulum path when it is hanging at its lowest point; the vector component increases as the pendulum is raised. More work is required to move it equal distances the farther it is raised.

CHECK QUESTION: Keeping the spring balance always perpendicular to the string, predict what the force will be if the string is pulled through 90 degrees and is horizontal? [The pull force will be equal and opposite to the force of gravity on the bob — its weight.]

Point out that the arc path to any elevation is longer than the vertical path. State that to compute the work done along the arc path is complicated because the force continually varies with distance. The computation requires a form of mathematics (integral calculus) that progressively adds a succession of tiny work segments. This advanced mathematics can be short-circuited, for the same answer is obtained by simply multiplying the weight of the bob by the vertical distance it is raised! The work done against gravity in elevating the pendulum bob is the same along either path, straight up or along the arc. Gravitational potential energy depends only on weight and height — not on the path taken to get it there. Discuss the elevated boulder in text Figure 8-3.

CHECK QUESTIONS: Does a car hoisted for lubrication in a service station have PE? [Yes, any elevated body has PE with respect to any chosen reference level, usually the "ground level."] How much work will raise the car twice as high? [Twice the work, also it will have twice as much PE.] How much work is required to raise it three times as high, and how much PE will it have? [Three times as much of each]

Kinetic Energy: A moving body can do work because of its motion. It has motion energy or, in Greek, kinetic energy (KE). Relate KE to force x distance. Discuss text Figure 8-5. State that later in the course you will apply the idea of KE to molecules and speak of heat and temperature. KE underlies heat (haphazard motion of molecules), sound (vibratory motion of molecules), and light (emitted by the vibratory motion of electrons in an atom).

CHECK QUESTIONS: Does a car moving along a road have KE? [Yes, any moving object has KE. KE is a relative quantity, as is speed. For example, the cup of tea you hold in a high-flying jet liner has KE with respect to the ground, but it has no KE with respect to the saucer in which it sits.] If the speed of the car doubles, by how much does the KE increase? [4 times] If the speed triples? [9 times]

(This is a good place to break.)

Conservation of Energy: When you rub two sticks together to start a fire, you transform mechanical energy to heat energy. When you do work and wind a spring in a toy cart, you give it PE, which then transforms to KE when the cart speeds up on the floor (text Figure 8-6). When the speed becomes constant, the continued transfer of PE is transformed to heat in order for the cart to work against friction. Without friction, KE would keep increasing with decreasing PE.

DEMONSTRATION [8-1]: Swing a pendulum to and fro and cite the transformations from PE to KE, and so on. Acknowledge the role of friction in damping the pendulum motion and the subsequent warming of the room!

Discuss text Figure 8-7.

DEMONSTRATION [8-2]: Preview electricity and magnetism and bring out the hand-cranked, horseshoe-magnet generator that lights up the lamp. Have student volunteers observe and state that more work is needed to turn the crank when the lamp is connected than when it is not. Then relate this to Think and Explain Question 7 in the text.

Discuss the question about the miracle car at the top of text page 109. Return to this question later when you discuss efficiency, and then repeat it for the case of a car that is 30 percent efficient to explain how the car will therefore go 30 percent of the 20 km distance.

Discuss the essentials for a follow through to energy conservation, use the experiments *Releasing Your Potential* or *Conserving Your Energy*.

(This is a good place to break.)

Machines: Apply energy conservation to the lever (Figure 8-9). Do not confuse the distances moved with the lever-arm distances of torque found later in Chapter 11. Here *Fd* refers to the force multiplied by the distance the "force moves" (parallel to the force), whereas in the case of torque *d* refers to the leverage distance that is perpendicular to the applied force. Show how varying the position of the fulcrum changes the relative values of output force and distance moved. Stress that this is in accordance with the rule, *work input = work output*.

CHECK QUESTION: Archimedes, the most famous scientist in ancient Greece, stated that if given a long enough lever, he could move the world. What is meant by this? [In accordance with the lever equation, $fD = Fd$, a force as great as the weight of the world could be lifted with the force he could muster, providing there was a place for the fulcrum!]

Acknowledge the different types of levers shown in text Figure 8-11 without overstating the distinction between the three types. The point to stress is the relationship $fD = Fd$, energy conservation.

CHECK QUESTION: In which type of lever is work output greater than work input? [NONE! In no system can work output exceed work input! Be clear about the difference between *work* and *force*.]

Show that a pulley is simply a lever in disguise, as shown in text figure 8 - 12. Show also $fD = Fd$.

DEMONSTRATION [8-3]: Use model pulleys to show the arrangements in text Figure 8-12. A spring balance will show the relative forces needed to support the same load.

CHECK QUESTION: In what pulley arrangement can work output exceed work input? [NONE!]

Cite the cases of charlatans who devise complicated arrangements of levers, pulleys, and other gadgets, such as magnets, to design a machine that will have a greater work output than work input. Such charlatans, very numerous in the past, are still at work today. They tell their "followers" that the oil companies and the patent office are conspiring against them, and that if they can raise sufficient funds, they can build their machine and usher in a better world. They prey on people who are ignorant of, or do not understand, the message of the energy conservation law. You can't get something for nothing. In fact you can't even break even, because of the inevitable transformation of energy to heat.

Efficiency: It should be enough that your students are acquainted with the idea of efficiency and actual and theoretical mechanical advantage. It is easy to let the plow blade sink deeper in text Section 8.8 and to turn this chapter toward the burdensome side of study. Therefore, I recommend this section be treated lightly and not be used as primary examination fodder.

CHECK QUESTION: What does it mean to say that a certain machine is 30% efficient? [It means it will convert 30% of the energy input to useful work — 70% of the energy input will be wasted.]

The efficiency of a light bulb underscores the idea of *useful energy*. To say an incandescent lamp is 10% efficient is to say that only 10% of the energy input is converted to the useful form of energy, light. All the rest goes to heat. However, even the light energy converts to heat upon absorption, so all the energy input to an incandescent lamp is converted to heat. This means it is a 100% efficient device as a *heater*, but not as a device for emitting light! We return to this idea in Chapter 24, *Thermodynamics*.

Fuels such as oil, gas, and wood are forms of concentrated energy. When they are used to do work, their energy is degraded. If all concentrations of energy are degraded, no more work can be done. Heat is the destiny of useful energy.

Section 8.9, *Energy for Life*, can be skimmed, or it can be the topic of a lecture on how the conservation of energy underlies all biology.

More Think-and-Explain Questions

1. Why does the force of gravity do no work on a bowling ball rolling along a bowling alley?

   Answer: Work is done on an object only when a force, or some component of a force, acts in the direction of motion of the object. In the case of the bowling ball, the force is perpendicular to the motion with no component parallel to the alley. (We will see later that, for the same reason, the force of gravity does no work on satellites in circular orbit.)

2. What effect does a long rifle barrel have on the muzzle velocity of the bullets it fires, and why?

   Answer: The long barrel provides a greater distance for the applied force of the expanding gases; so more work is done on the bullet, and it emerges with greater velocity and greater KE. (Of course there is a limit to the rifle barrel's length. Expanding gases would dissipate in a barrel that is too long and friction would prevail.)

3. Can an object have mechanical energy without having momentum? Explain. Can an object have momentum without having energy? Explain.

   Answer: First of all, all objects have internal heat energy. (They also have energy of being in the form of mass, $E_O = mc^2$.) If the mechanical energy is due to motion, then a moving object has both KE and momentum. If the mechanical energy is PE and the object is at rest, then it can have mechanical energy without having momentum. But if an object has momentum, then, by definition, it is moving and also has KE.

4. Why are there long handles and short blades on metal-cutting shears? (This is just the opposite for ordinary scissors.)

   Answer: The long handles act as long levers and multiply the force applied to the blades. Such an increased force is not needed for ordinary scissors, which have much shorter handles.

5. Suppose you are at the edge of a cliff, and you throw one ball down to the ground below and another one up at the same speed. The upwardly-thrown ball rises and then falls to the ground below. How do the speeds of the balls compare when they strike the ground? Neglect air resistance and use the conservation of energy to arrive at your answer.

   Answer: The ball strikes the ground with the *same* speed, whether thrown up or down. Relative to the ground below, each ball starts with the same energy (PE + KE). When they reach ground level PE = 0, all the energy is the same KE. This can be seen by the analysis of Figure 2-6 on text page 18, where the ball that is thrown upward will return to its starting level at its launching speed. Its speed at the ground is the same whether the ball is thrown up or down. (Interestingly enough, the speed of the ball upon striking the ground will be the same if the ball is thrown horizontally, or in any direction! Velocity will be different because of different directions.)

Computational Problems

1. With respect to the ground below, how many joules of PE does a 1000-N boulder have at the top of a 5-m ledge? If it falls, with how much KE will it strike the ground? What will be its speed of impact?

   Answer: PE = weight x height = 1000 N x 5 m = 5000 Nm = 5000 J. By the conservation of energy, the boulder will have the same 5000 J of KE at impact. It will have a speed of 10 m/s upon impact, just as any object will freely fall a distance of 5 m during its first second of fall (Chapter 2), and any object starting from rest will reach a speed of about 10m/s in its first second of fall (actually 9.8 m/s). Or, from KE = PE, $1/2\ mv^2 = mgh$, or $v^2 = 2gh$, or $v = \sqrt{2gh} = \sqrt{2(10)(5)} = \sqrt{100} = 10$ m/s.

2. Use the conservation of energy to find an equation for the speed of a freely-falling object that falls from rest at a height h. That is, equate the PE to KE and solve for velocity v.

   Answer: $v = \sqrt{2gh}$: The PE (mgh) of the object at the rest position will be converted to KE ($1/2\ mv^2$) after falling through a vertical distance h. From $mgh = 1/2\ mv^2$, we see after cancelling m and rearranging terms, $v = \sqrt{2gh}$.

3. Water drops about 50 m over the Niagra Falls. If 8 million kg of water fall each second, what power is available at the bottom?

   Answer: 4000 MW: $P = W/t = mgh/t =$ (8 000 000 x 10 x 50) J/s = 4 000 000 000 W, or 4000 MW.

4. When a 1-kg projectile is fired at 10 m/s from a 10-kg gun, the momentum of the projectile and recoiling gun is the same, but in opposite directions. Show that the KE of the projectile is 10 times greater than the KE of the recoiling gun.

   Answer: KE of projectile = $1/2\ mv^2$ = 1/2 (1 kg) (10 m/s)^2 = 50 J. The speed of the recoiling gun, by momentum conservation, must be 1 m/s. The KE of the recoiling gun = $1/2\ mv^2$ = 1/2 (10 kg) (1 m/s)^2 = 5 J, hence the KE of the projectile is 10 times the KE of the recoiling gun.

5. What is the theoretical mechanical advantage of a 5-m long inclined plane with one end elevated to 1 m? What is the actual mechanical advantage if 100 N of effort is needed to push a 400-N block of ice up the plane? What is the efficiency of the inclined plane?

   Answers: TMA = (Input distance)/(output distance) = 5 m/1 m = 5 AMA = (Output force)/(input force) = 400 N/100 N = 4. Efficiency = AMA/TMA = 4/5 = 0.8, or 80%.

6. What is the efficiency of the body when a cyclist expends 1000 watts of power to deliver mechanical energy to the bicycle at the rate of 100 watts?

   Answer: Efficiency = Power output/power input = 100 W/ 1000 W = 0.10, or 10%.

9 Circular Motion

To use this planning guide work from left to right and top to bottom.

Chapter 9 Planning Guide
• The bulleted items are key: Be sure to do them!

Topic	Exploration	Concept Development	Application
Rotation		• Text 9.1–9.2/Lecture Demos 9-1, 9-2	
Centripetal and Centrifugal Force	Act 30 (1 period)	• Text 9.3–9.5/Lecture Demos 9-3, 9-4	Nx-Time Q 9-1
Simulated Gravity		• Text 9.6/Lecture Con Dev Pract Pg 9-1	

Video: *None*
Evaluation: Chapter 9 Test

See the Chapter Notes for alternative ways to use all these resources.

Objectives

After studying Chapter 9, students will be able to:
• Distinguish between rotate and revolve.
• Distinguish between linear speed and rotational speed and explain what each depends on.
• Give examples of a centripetal force.
• Describe the resulting motion of an object if the centripetal force acting on it ceases.
• Explain why it is incorrect to say that a centrifugal force pulls outward on an object being whirled in a circle.
• Explain why a ladybug in a whirling can would experience what seems to be an outward force, and why a physicist would label this a fictitious force.
• Describe how a simulated gravitational acceleration can be produced in a space station.

Possible Misconceptions to Correct

• Linear speed and rotational speed are the same.
• The linear speed on a rotating surface is the same at all radial distances.
• Things moving in a circular path are pulled outward by some force.
• If the string breaks on an object pulled into a circular path, the object will move radially outward (rather than tangentially).

Demonstration Equipment

• [9-1] Two coins and a rotating turntable of any kind
• [9-2] Meter stick (or any stick) to swing as a pendulum
• [9-3] A tin can (or similar object to whirl) tied to a piece of string about 1 m long
• [9-4] Bucket of water to swing overhead

Introduction

This chapter extends the translational ideas of Chapter 2 to rotation and serves as an introduction to Chapter 11. A space station theme is briefly treated to open the door to the myriad of interesting physics applications in a rotating frame of reference.

The classic oldie-but-goodie PSSC film *Frames of Reference* goes nicely with this chapter. For good ideas read "Which Way is Up" by Al Bartlett, *The Physics Teacher*, No. 10, p. 429 (1972); "Up in the Lab and in Literature" by D. Easton, *The Physics Teacher*, No. 22, p. 100 (1984); and the not-yet-outdated books *The High Frontier* by

Gerard O'Neill (Morrow, 1976); and *The Third Industrial Revolution* by Harry Stine (Ace, 1975).

Note that this chapter omits entirely the "right hand rule," where fingers of the right hand represent the motion of a rotating body and the thumb represents the positive vector of motion. I have always felt that the reason for this rule in introductory physics courses was to provide the instructor the opportunity to easily make up tricky exam questions. Please spare giving undue emphasis to this material, which your students can get into in a follow-up course.

It is often said that rotational motion is analogous to linear motion and therefore it should not be difficult to learn. But consider the many distinctions between 1) linear, 2) rotational, 3) revolutional, 4) radial, 5) tangential, 6) angular motion and that all are related to a) speed, b) velocity, and c) acceleration. We also have centripetal and centrifugal forces, real and ficticious, not to mention torques that will be coming up soon. So there is a myriad of ideas to describe and understand. Is it any wonder why students who do well in linear motion fall apart here? A study of rotational motion is considerably more complex than a study of linear motion. Caution your students to be patient with themselves if they don't immediately catch on to what has taken centuries to master.

Suggested Lecture

Rotations and Revolutions: Distinguish between a rotation (spin about an axis located within a body) and a revolution (axis outside the body). A wheel rotates; its rim revolves. Cite the case of a spinning satellite (rotating) while it orbits the earth (revolving).

> CHECK QUESTIONS: Does a tossed football rotate or revolve? [It rotates (spins).] Does a ball whirled overhead at the end of a string rotate or revolve? [It revolves around you.]

Rotational Speed: Distinguish between rotational speed and linear speed by the examples of riding at different radial positions on a merry-go-round.

> DEMONSTRATION [9-1]: Place coins on the top of a turntable, one near the center and the other near the edge. Rotate the table and show how the outer coin has a greater linear speed. Point out that both coins have the same rotational speed — they undergo the same number of revolutions per second.

Compare the speeds of the coins to the speed of different parts of a phonograph record beneath the stylus. Linear velocity that is perpendicular to

the radial direction is the same as tangential velocity [$v = r\omega$]. Cite examples such as being able to see detail on the hub cap of a moving car while not seeing similar detail on the tire. Cite the motion of "tail-end Charlie" at the skating rink.

> DEMONSTRATION [9-2]: Let a meter stick supported at the 0-cm mark swing like a pendulum from your fingers. How fast at any given moment is the 100-cm mark moving compared to the 50-cm mark? [The 100-cm mark is twice as far from the center of rotation than the 50-cm mark is, and it has twice the linear speed.]

Liken this idea to the speed of the end of a fly swatter. (The relatively long handle greatly amplifies the speed of your hand.)

Centripetal Force: Define centripetal force as any force that causes a body to move in a circular path — or in part of a circular path, such as rounding a corner while riding a bicycle.

> DEMONSTRATION [9-3]: Whirl an object, such as a tin can securely fastened at the end of a string, above your head. Don't simply discuss this. DO IT!

Expand on the idea that a centripetal force is exerted on the whirling can at the end of a string. The string pulls radially inward on the can. By Newton's third Law, the can pulls outward on the string — so there is an outward-acting force on the string. Stress the fact that this outward force does not act on the can. Only an inward force acts on the can.

> DEMONSTRATION [9-4]: Swing a bucket of water in a vertical circle and show that the water doesn't spill. All your students have heard of this demonstration, but only a few have actually seen it done. They will particularly enjoy seeing YOU do it and the prospect of seeing you "all wet!" (You're copping out if you only talk about it. DO IT!)

The water doesn't spill at the top when the centripetal force is at least equal to the weight of the water. But before invoking this explanation, ask why the water doesn't fall. After some thought state that your question is misleading, for the water indeed does fall! The "trick" is to pull the bucket down as fast as the water falls so that both fall the same vertical distance in the same amount of time. (Too slow a swing produces a wet teacher.) Interestingly enough, the water in the swinging bucket is analogous to the orbiting of a satellite.

Both the swinging water and a satellite, such as the orbiting space shuttle, are falling. Because of their tangential velocities, they fall in a curve; at just the right speed for the water in the bucket and at just the right greater speed for the space shuttle. Tying these related ideas together is good teaching!

CHECK QUESTION: A motorcycle runs on the inside of a bowl-shaped track (see sketch). Is the force that holds the motorcycle in a circular path an inward- or outward-directed force? [It is an inward-directed force — a centripetal force. An outward-directed force acts on the inner wall, which may bulge as a result, but no outward-directed force acts on the motorcycle.]

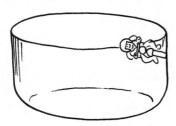

Centrifugal Force: Centrifugal force is the name given to a radially outward-acting force, and it is useful only in a rotating frame of reference. The inward push feels like an outward pull to the occupants in a rotating system, as if a big mass were out there causing gravity. State how it differs from a real force in that there is no interaction — that is, there is no mass out there pulling on it. The magnetic force between a paper clip and a magnet, for example, requires both the clip and magnet. They pull on each other and comprise a real interaction. Whereas a real force is an interaction between one body and another, there is no reaction counterpart to the centrifugal force that is felt. Distinguish centrifugal force from the action-reaction pairs of forces at the feet of an astronaut in a rotating station.

CHECK QUESTION: A person in a spinning space station feels a force like that of gravity and does push-ups against the floor just as she would at home. Is the force that is felt a centrifugal or centripetal force? [Centrifugal force] How does it differ from gravity? [Unlike gravity, there is no interaction with a "mass" or other thing that supposedly does the pulling.]

So we see that the centrifugal force is not a real force, but it is the effect of rotation. This distinction may be difficult for your students to make.

Applications: Discuss rotating space stations. Show how g varies with the radial distance from the hub and with the rotational rate of the structure.

More Think-and-Explain Questions

1. Somebody says that the reason the moon doesn't fall to earth is because the centrifugal force that acts on it exactly counteracts gravity. What do you say about this?

   Answer: No, no, no! First, no centrifugal force acts on the moon. The only force that holds it in orbit around the earth is a centripetal force, directed toward (not away from) the earth. This centripetal force is the gravitational force. Second, the moon does fall toward the earth. Because of its tangential velocity, it falls around, rather than into, the earth.

2. Is the centripetal force that holds the moon in earth orbit equal to the gravitational force?

   Answer: The centripetal force that holds the moon in an earth orbit is the gravitional force, so of course it is exactly the same in magnitude and in every other respect. It is a mistake to assume there is both a gravitational *and* a centripetal force acting on the moon.

Computational Problems

1. From the equation $F = mv^2/r$, calculate the tension in a 2-m length of string that whirls a 1-kg mass at 2 m/s in a horizontal circle.

   Answer: $F = mv^2/r = [1 \text{ kg} \times (2 \text{ m/s})^2] / 2 \text{ m} = 2 \text{ N}$.

2. Answer the previous question for the case of (a) twice the mass, (b) twice the speed, (c) twice the length of string (radial distance), and (d) twice mass, twice speed, and twice distance all at the same time.

   Answers: (a) Twice, 4 N; (b) four times, 8 N; (c) half, 1 N; (d) twice, 4 N (because 2 x 4 x 1/2 = 4 N).

10 Center of Gravity

To use this planning guide work from left to right and top to bottom.

Chapter 10 Planning Guide

• *The bulleted items are key: Be sure to do them!*

Topic	Exploration	Concept Development	Application
Center of Gravity		• Text 10.1–10.3/Lecture • Demos 10-1 to 10-3	
Toppling/ Stability		• Text 10.4–10.5/Lecture Demos 10-4, 10-5	Nx-Time Qs 10-1, 10-2
CG of People	Act 31 (1 period)	• Text 10.6/Lecture Demos 10-6, 10-7 Con Dev Pract Pg 10-1	

Video: *Center of Gravity*
Evaluation: Chapter 10 Test

See the Chapter Notes for alternative ways to use all these resources.

Objectives

After studying Chapter 10, students will be able to:
• Describe center of gravity.
• Describe how a plumb line and bob can be used to find the center of gravity of an irregularly-shaped object.
• Given the location of the center of gravity and the area of support of an object, predict whether the object will topple.
• Distinguish among stable equilibrium, unstable equilibrium, and neutral equilibrium.
• Give examples of how a human is affected by the need to keep the body's center of gravity over the support base.

Possible Misconceptions to Correct

• The center of mass of an object must be where its physical mass exists.

• The center of gravity of a person is at a fixed place inside the body.

Demonstration Equipment

• [10-1] Small ball and a piece of wood or plastic cut into an L shape about the size of this book, so it can be clearly seen
• [10-2] An irregularly-shaped piece of plywood somewhat larger than the size of this book, with 3 strings fastened to different locations along the edge
• [10-3] Same L-shaped object, a baseball bat, and any other irregular shapes to show CG
• [10-4] Any assortment of CG toys, see text Figure 10-18
• [10-5] A container of dried beans and a Ping-Pong ball, see text Figure 10-20
• [10-6] A chair

Introduction

Toppling is treated in this chapter before the concept of torque is introduced. Toppling is treated as a result of a nonsupported center of gravity (CG). I suggest you treat it as the text does and save the concept "torque" until the following Chapter 11, *Rotational Mechanics*. Or else you can introduce torque when you discuss toppling, as sort of a preview to the next chapter.

A nice extension of Activity 3 on text page 143 (sliding two fingers to the center of a balanced meterstick) is to do the same thing to a plate, which is horizontally-balanced on three fingers. Bring your fingers together and the plate remains in balance!

A nice prop is a cutout of your state, suspended as the cutout of the U.S. is in text Figure 10-8. (For contiguous U.S., the CG is near Lebanon, Kansas. For the entire U.S., including Alaska and Hawaii, the CG is in South Dakota west of Castle Rock.) You can use this prop to show where the CG of your homestate is. You can go a step further and stick nails or other heavy weights where population densities are greatest and, thus find the "CG" of your state in terms of population. (In 1980 the "center of population" for the United States was 1/4 mile west of De Soto, Mo.) Construction makes a good project for energetic students.

Another good prop is the variable incline, block, and plumb bob arrangement shown in text Figure 10-10 — another student project!

An impressive demonstration is the following: Place a wind-up (spring action) stop watch on top of an upside-down watch glass with a tiny mirror siliconed to the watch. Shine a laser beam on the mirror and watch the beam move back and forth with the rocking motion of the watch. The CG is slightly displaced with each tick! The same is true of a person lying still on a table. The table vibrates slightly with the heart beat.

This chapter can be skimmed or skipped if your time is tight.

Suggested Lecture

Begin with the following demonstration.

DEMONSTRATION [10-1]: Toss a small ball from one hand to the other and call attention to the smooth parabola it traces. Then toss an L-shaped piece of wood or other material.

State that in doing so it doesn't seem to follow a smooth parabola. It wobbles all over the place; very special place nonetheless — the place discussed in this chapter — the CG.

Center of Gravity and Center of Mass: Define the center of gravity as the average positon of weight, and define the center of mass as the average position of matter. For our purposes, they describe the same point. Show that the CG of a book is at the geometrical center, which is easily found by the intersection of diagonal lines from opposite corners. But what of an object with more than 4 corners?

DEMONSTRATION [10-2]: Show a piece of irregular-shaped plywood (somewhat larger than the book) that has about 5 corners. Three short pieces of string are fastened at different places along its edge. Ask how the CG of this shape can be found. Certainly it's not by connecting the corners, because the lines would not have a common intersection. Solve this by suspending the plywood by one of the strings and draw a vertical chalk line beneath the point of suspension. Do this with a second string and the intersection is the CG. To double check, suspend it by a third string.

DEMONSTRATION [10-3]: Try balancing the L-shape with its small end on the table and watch it topple. That's because it wasn't supported at its "average position of weight" — its CG. Illustrate CG with a baseball bat and other objects.

State that when the "average position of mass" is considered, one speaks of the center of mass. For most cases the two are indistinguishable, so CG will be taken to mean both.

Stability: Distinguish between unstable, neutral, and stable equilibrium.

DEMONSTRATION [10-4]: Show how the CG of an object is either raised, not changed, or lowered when the object is tipped. Also demonstrate, if available, the devices shown in text Figures 10-17 and 10-18.

Work and Raising CG: A floating iceberg will not tip over because its CG is below the water line. If it were to tip, its CG would be raised, which requires a work input. This is also true for Sutro Tower in San Francisco which easily withstands strong winds. Its base is so heavily buried into concrete that in a sense it is "already tipped over." The same is true for the Space

Needle in Seattle. Work input is needed to raise the CG of a system. Relate this idea to text Figures 10-20 and 10-21.

DEMONSTRATION [10-5]: Shake a container of dried beans with a Ping-Pong ball at the bottom, as shown in text Figure 10-20. Here the density of the beans is greater than the density of the Ping-Pong ball, so it's like "panning gold." For objects of the same density, smaller ones will fill in the open spaces between larger ones to produce a greater effective density of space. The CG lowers.

The bean-shaking demonstration can be extended to the Ping-Pong ball in a glass of water. The CG of the system is lowest when the Ping-Pong ball floats. Push it under the surface and the CG is raised. If you do the same thing with something more dense than water, the CG is lowest when it is sunk at the bottom (more about this in Chapter 19).

CG of People: Ask your students if they have a CG. Acknowledge that the CG in men is generally higher than it is in women (1% - 2%). This occurs mainly because women tend to be proportionally smaller in the upper body and heavier in the pelvis. The CG of men and women can be likened to the CG of a baseball bat standing on end. With its heavy end toward the ground, it is similar to the build of an average woman. With its narrow end toward the ground, the bat is similar to the build of an average man. Point out that this may apply to most people, but not to all people. The CG of most adults, when their arms are at their sides, lies about 6 inches above the crotch and a bit below the belly button. (If your students have done the lab activity *Where's Your Center of Gravity*, they have presumably discovered this.) In children, the CG is about 5% higher because of their proportionally larger heads and shorter legs. The central location of the CG is important in a fetus, because unborns rotate. The CG of a person is not located in a fixed place, since CG depends on body orientation. Just as the CG of a boomerang is outside the material, your CG is outside your body when you bend over and make a U or L shape. Whatever your body orientation, in order to remain stable, your CG must be above (or below) a support base area. Show how this support base is enlarged when you stand with your feet wide apart (see text Figure 10-23).

DEMONSTRATION [10-6]: Attempt to stand from a seated postion without putting your feet under the chair. This cannot be done unless your feet provide a support base area that lies beneath your CG.

DEMONSTRATION [10-7]: Challenge your students to do the following home project. Tell them to stand facing a wall with toes against the wall and simply stand unaided on tiptoes for a couple of seconds. If they can do this in your next class meeting, they will receive an A for the course! (This cannot be done, because the CG will be farther from the wall than the support base provided by the toes. With a bit of "trickery", this can be done near a doorway when you hold a heavy weight from your extended arm into the doorway in front of yourself so that the CG is above the narrow support base. But unaided, you'll award no A grades for this feat.)

More Think-and-Explain Questions

1. Why is it dangerous to roll open the top drawers of a fully-loaded file cabinet that is not secured to the floor?

   Answer: The CG of the cabinet may extend beyond its base and topple.

2. Why is the middle seating most comfortable in a bus traveling along a bumpy road?

   Answer: The bus will tend to rock about its CG. The closer one's seat is to the CG, the less up and down motion one experiences as the bus rocks. (The CG of the bus will depend also on whether the motor is in the front or the back.)

3. What is the role of the heavy tail of dinosaurs with respect to CG?

   Answer: With its tail extended, the dinosaur's CG is above its feet when it runs.

4. How does a heavy tail enable a monkey standing on a branch to reach to farther branches?

   Answer: With tail extended backward, the monkey's CG is above its feet even when its arms are extended forward.

5. The CG of a baseball bat is not in the middle of the bat, but toward the more massive end. When the bat is stood upright with its massive end down, its CG is lower to the ground than when it is stood upright with its massive end up. How is this similar to the different CGs of broad-shouldered men and broad-hipped women?

   Answer: Standing upright with the massive end up, the bat's CG is similar to the higher CG of the man; standing with the massive end down, it is similar to the lower CG of the woman.

To use this planning guide work from left to right and top to bottom.

Chapter 11 Planning Guide
• The bulleted items are key: Be sure to do them!

Topic	Exploration	Concept Development	Application
Torque	• Act 32 (1 period)	• Text 11.1–11.3/Lecture • Demos 11-1–11-5 Exp 33 (>1 period) Con Dev Pract Pg 11-1	Exp 34 (1 period) Nx-Time Qs 11-1, 11.2
Rotational Inertia	Act 35 (1 period)	• Text 11.4–11.5/Lecture Demos 11-6 – 11-9	
Angular Momentum		• Text 11.6–11.7/Lecture Demos 11-10, 11-11	

Video: *Rotational Mechanics*
Evaluation: Chapter 11 Test

See the Chapter Notes for alternative ways to use all these resources.

Objectives

After studying Chapter 11, students will be able to:

• Define torque and describe what it depends on.
• Describe the conditions required for one torque to balance another.
• Given the location of the center of gravity of an object and the position and direction of the forces on it, tell whether the forces will produce rotation.
• Describe what the rotational inertia of an object depends on.
• Give examples of how a gymnast changes the rotational inertia of her body in order to change the spin rate.
• Define angular momentum and describe the conditions under which it (a) remains the same and (b) changes.
• Give an example of a situation in which rotational speed changes but angular momentum does not.

Possible Misconceptions to Correct

• Torque and force are the same concept.
• Inertia and rotational inertia are much the same concept.

Demonstration Equipment

• [11-1] L-shape used in Chapter 10 demonstration
• [11-2] Your own body
• [11-3] Disk with lead set in it so that the CG is off center and an inclined plane
• [11-4] Two drinking glasses, a candle, and two pins or needles
• [11-5] A very heavy plank which has a weight equal to a significant fraction of your own
• [11-6] A wooden stick about 1 m long with a 1-kg or 2-kg mass attached at one end, or substitute a broom or sledge hammer
• [11-7] A pair of plastic pipes about 1 m long; one with 2 massive lead inserts in its center and the other with massive lead inserts in each end (hidden version of text Figure 11-10)
• [11-8] Hoop, solid cylinder, and inclined plane (a board 1 meter or longer will do)
• [11-9] Cans of food, filled with liquids and solids, and an inclined plane
• [11-10] Platform to stand on that will rotate freely, two weights
• [11-11] Platform to stand on that will rotate freely, a bicycle wheel

Introduction

This chapter may be difficult for your students, because it has a lot of hard core physics that takes time to assimilate. It can be omitted without consequence.

Chapter 10, *Center of Gravity*, may be combined with this chapter. In Chapter 10, toppling was explained in terms of the CG of a system extending beyond a base of support, and torques were not discussed. The concept of torque is introduced here.

The formulas for the rotational inertia of common shapes are given in Figure 11-15 on text page 152. These formulas are intended to serve as interesting information about relative rotational inertias. For some of your students, however, they may be seen as a threat, particularly if they are considered for closed book exams. It would be counterproductive for your students to learn the different formulas, or for them to think they are learning any physics at all if the formulas are memorized. Please make this clear to your class, and do not use the formulas as examination material.

An excellent resource for examples of rotational mechanics is *Sport Science* by Peter Brancazio (Simon & Shuster, 1984). Brancazio's book is an excellent source of physics applications found in the human body.

Suggested Lecture

Torque: Extend the activity *Torque Feeler* to other "twisting forces," such as prying a lid off a can with a screwdriver, turning a wrench, or opening (rotating) a door. Cite how a steering wheel is simply a modified wrench, and explain why trucks and heavy vehicles without power steering use large-diameter steering wheels. In all these cases there are two important considerations, (1) the application of a force and (2) leverage.

> DEMONSTRATION [11-1]: Place the L-shaped object on the table and show how it topples in different positions, just as you did in Demonstration [10-2] in the previous chapter.

On the board draw the L-shape as shown in the first sketch and define torque.

Torque and Your CG: Recall from the previous chapter that the location of one's CG depends upon body configuration.

> DEMONSTRATION [11-2]: Stand against a wall and ask if it is possible for you to bend over and touch their toes without toppling forward. When you attempt to do so, your body position

approximates the L-shape. Sketch this on the board next to the L-shape as shown. By now your chalkboard looks like the sketch below.

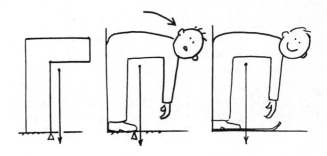

Discuss a remedy for such toppling, for example wearing longer shoes, snowshoes, or skis. Sketch a pair of skis on the feet of the person in your drawing.

The Pregnant Woman: Change the example and ask why a pregnant woman often gets back pains. Sketch a woman in the very early and very late stages of pregnancy to how the CG shifts forward — beyond a point of support for the same posture. Make a third sketch showing how the woman adjusts her posture so that her CG is above the support base bounded by her feet. Draw in the "marks of pain." Ask the class what the woman could do to relieve these pains. If someone in class doesn't volunteer the idea of her wearing skis, do so yourself and sketch skis on her feet in the second drawing.

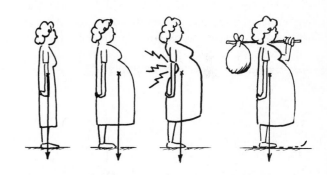

Suggest an alternate solution to your class — a pole, with a load near the end, carried on the woman's shoulder. Erase the skis and sketch in the pole and load as shown. Acknowledge that the woman would have to increase the mass of the load as the months go by and ask what could be done instead. Someone should volunteer that she need only move the load closer to the end of the pole, which in effect shifts the overall CG in a favorable direction (or produces more counter torque). This routine is effective and sparks much class interest. You may, however, prefer to develop a similar story using a man with a paunchy stomach.

Torque and CG Examples: Return to your chalkboard sketches of L-shaped objects and relate their tipping to the torques that exist. Point out the lever arms in the sketches.

CHECK QUESTION: An L-shaped object with CG marked by the X rests on a hill as shown. Draw this on your paper and mark it appropriately to determine whether the object will topple or not. (Be prepared for some students to incorrectly sketch in the "vertical" line through the CG *perpendicular* to the slope as shown. Vertical lines should be vertical!)

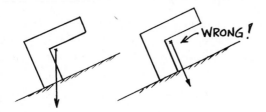

Cite some examples involving the CG in animals and people: a walking pigeon moves its head back and forth with each step, thereby keeping its CG above the foot it momentarily stands on; the long tails of monkeys enable them to lean forward without losing balance; people lean backwards when carrying a heavy load at their chests; and the effective method for carrying a heavy load by dividing it into two parts and suspending one part at each end of a yoke supported in the center by a man's shoulders.

Ask why a ball rolls down a hill. State that "because of gravity" is an incomplete answer. Gravity would have it slide down the hill. The fact that it rolls, or rotates, is evidence of an unbalanced torque. Sketch this on your chalkboard.

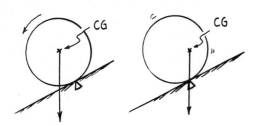

DEMONSTRATION [11-3]: Show how a "loaded disk" rolls *up* an inclined plane. After class speculation, show how the disk remains at rest on the incline. Modify your chalkboard sketch to show the CG directly above the support point. This means there is no lever arm and therefore no torque.

Seesaws: Extend the discussion of rotation to seesaws. Show this first on the board with a sketch of equal-weight players (a sketch similar to

Figure 11-5 on text page 147). Cite how the force times the distance on the left side of the fulcrum that tends to make the seesaw rotate counterclockwise equals the force times the distance on the right side of the fulcrum that tends to make it rotate clockwise. Then discuss the case of the twice-as-heavy boy. (Figure 11-5 on text page 147 uses the same idea.)

CHECK QUESTION: If the boy in Figure 11-5 weighed 600 N, how far would he have to sit from the fulcrum for equilibrium to occur? [1 m.]

Possible point of confusion: Here we say Torque = Fd and in the previous chapter we said Work = Fd. The distance for torque is altogether different than the distance used for work. In work, the distance d is the distance the "force moves" (parallel to the force). In torque, the distance d refers to the leverage distance (perpendicular to the force). The ratios d/d for levers and seesaws, however, turn out to be the same.

Explain how participants on a seesaw can vary the net torque by not only sliding back and forth, but also by leaning. In this way the location of their CGs, and hence the lever arm distance, is changed.

DEMONSTRATION [11-4]: Do as Cindy Dube at Farmington High School in Connecticut does and make a candle seesaw. Trim the candle wick to expose both ends, then balance the candle on a needle pushed through the center. Rest the ends of the needle on a pair of drinking glasses. Light both ends of the candle. As the wax drips, the CG shifts and causes the candle to oscillate.

Solitary Seesaw: Thus far, the CG of the seesaw has been at the axis of rotation. The weight of the seesaw contributed no torque because of the zero lever arm. When the fulcrum is not at the CG, the situation becomes somewhat more complicated, and more interesting.

Cite the case of the obnoxious boy who cannot find a playmate to join him playing seesaw. He simply moves the middle of the seesaw beyond the fulcrum and uses the weight of the seesaw at its CG as his "invisible partner." Once balanced, he is able to rotate up and down by leaning toward and away from the fulcrum. This "solitary seesaw" idea will require repeated explanation. It is the concept treated in the lab experiment, *Weighing an Elephant*.

DEMONSTRATION [11-5]: Place a heavy plank on your lecture table so that it hangs over the edge. Walk out on the overhanging part and ask why you don't topple. (You can simulate this with a smaller model. Use a heavy weight and a board hanging over the edge of the table.) Relate this setup to the solitary seesaw.

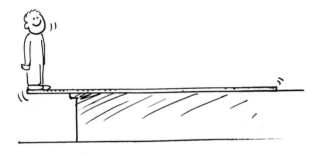

This demonstration is difficult for students who don't see it as a version of the solitary seesaw. If the person and the board have the same weight, then the fulcrum is midway between the CGs of both, like the simpler case of twins on a seesaw. Likewise, if you and the board are equal in weight, the edge of the table will be midway between you and the CG of the board — at the 1/4 mark (one half the distance from the end of the board to its CG). If you are heavier than the board, then the fulcrum (edge of the table) must be closer to you so that the board's CG has more leverage.

(This is a good place to break.)

Rotational Inertia: Compare the concept of inertia and its role in linear motion to rotational inertia (sometimes called "moment of inertia") and its role in rotational motion. The difference between the two involves the role of *radial distance* from a rotational axis. The greater the distance of mass concentration is, the greater is the resistance to rotation.

DEMONSTRATION [11-6]: Have students try to balance an long upright stick with a massive lead weight at one end on one finger. Try it first with the weight at the finger tip, then try it with the weight at the top. [Rotional

inertia is greater for the stick when it is made to rotate with the massive part far from the pivot. In other words, the farther the mass is from the pivot, the greater the rotational inertia is — which is to say, the more it resists a change in rotation.] You can substitute a broom or long-handled hammer. Relate this to the ease with which a circus performer balances a pole laden with people doing acrobatics, and cite how much more difficult it would be for the performer to balance an empty pole!

Relate this demonstration, and the continuous adjustments needed to keep the object balanced, to the similar adjustments that are necessary for keeping a rocket vertical when it is first fired. Amazing!

DEMONSTRATION [11-7]: You need two 1-meter pipes, one with two lead plugs in the center and the other with a lead plug in each end. They appear identical. Weigh them to show that they are the same weight. Give the pipe with a plug in each end to a student and ask her to rotate it about its center (as in text Figure 11-10). Have another student do the same with the pipe with the plugs in the middle. Then have them switch pipes. Good fun. Next ask the students to speculation as to why one was noticeably more difficult to rotate back and forth.

Explain how the location of an object's mass with respect to its axis of rotation determines its rotational inertia (hence the large rotational inertias of flywheels with mass concentrations along the rim). The rotational inertia of an object is a measure of how much it resists turning, a fact employed by tightrope walkers who carry long poles.

Cite the similarity of training wheels on a beginner's bicycle to the long pole used by a beginning tightrope walker. Suppose the ends of the long pole could slide along supporting slots as a tightrope walker walked along the wire. If the pole has adequate rotational inertia, the slots provide psychological comfort as well as actual safety. Just as the training wheels could be safely removed without the rider's knowledge, the slots could be safely removed without the walker's knowledge. Angular momentum aids the cyclist and rotational inertia aids the tightrope walker.

State the law of rotational inertia: An object rotating about an axis tends to keep rotating in the absence of an external torque.

Ask why a football is set spinning when it is thrown. Show also for minimum air resistance and maximum range that a second spin is imparted about another axis; a very slow spin that finds the nose of the ball always aligned with its trajectory.

CHECK QUESTION: Why are spiral grooves cut in the bore of a gun? [Once spinning about an axis, the bullet tends to remain spinning about that axis. This prevents a flip-flop by the bullet and avoids greater air resistance.]

Show how a longer pendulum has a greater period and relate this to the different strides taken by people with long or short legs. Imitate these strides yourself — or at least show this with your fingers walking across the desk.

Discuss the variety of rotational inertias shown in Figure 11-15 on text page 152. Stress that the formulas are for comparison and point out why the same formula applies to the pendulum and the hoop. [All the mass of each is at the same distance from the rotational axis.] State how reasonable the smaller value for a solid disk is, given that much of its mass is close to the rotational axis. State that learning the formulas has little value. Comparing the effort needed to change the rotation of the various figures, on the other hand, is good physics.

CHECK QUESTION: Suppose the shapes in text Figure 11-15 all have the same mass and radius (or length). Which would be the easiest to start spinning about the axes indicated? [The stick about its CG, as indicated by its small rotational inertia] Which would be the most difficult to start (or stop) spinning? [The simple pendulum or the hoop about its normal axis, as indicated by the relatively large rotational inertia]

Relate the acceleration of a rolling object down an incline to its rotational inertia. For similar masses, shapes with the greater rotational inertia (the "laziest") lag behind the shapes with less rotational inertia.

DEMONSTRATION [11-8]: Roll hoops and cylinders down an incline after asking your students to predict which rolls faster? (This is the lab activity *Rotational Derby*.) [Any cylinder will beat any hoop. Mass does not play a role, just as it doesn't play a role in the acceleration of free fall or the acceleration of a block down a frictionless plane. The acceleration down an incline has to do with the rotational inertia *per kilogram*. In effect, the mass cancels out. That's why, regardless of mass, any cylinder will beat any hoop.]

DEMONSTRATION [11-9]: Roll various cans of food down an incline and ask why different cans roll at different accelerations. Does a can of chicken soup beat vegetarian vegetable? How do Chinese noodles compare with pineapple juice? [A can of low-viscosity liquid

rolls faster than a can of beans — the beans are made to turn with the turning can and are therefore "lazy," while a liquid doesn't turn with the can and behaves like a sliding block — no rotation, no rotational inertia! Contents of various compositions and viscosities will produce varying performances. (Again, this is the object of *Rotational Derby*).]

Going Further with Rolling: Rolling things have two kinds of kinetic energy, that due to linear motion and that due to rotational motion. An object rolling down an incline will lag behind a freely-sliding object, because a part of a rolling object's kinetic energy is in rotation. If this is clear, the following question is in order for your better students.

NEXT-TIME QUESTIONS: Which will roll with greater acceleration down an incline, a can of water or a frozen can of ice? Give double credit for a good explanation of what is seen. [The can of water will undergo appreciably more acceleration, because the liquid is not made to rotate with the rotating can. It, in effect, "slides" rather than rolls down the incline, so practically all the KE at the bottom is in linear speed with next to none in rotation. Fine, one might say, then if the liquid doesn't rotate, the can ought to behave like an empty can with the larger rotational inertia of a "hoop" and lag behind. This brings up an interesting point. The issue is not which can has the greater rotational inertia, but rather which can has the greater rotational inertia compared to its mass (note the qualifier in the legend of text Figure 11-16.) The liquid content has appreciably more mass than the can that contains it, hence the non-rolling liquid serves to increase the mass of the can without contributing to its rotational inertia. The can of liquid has a relatively small rotational inertia compared to its mass.]

Rotational Inertia and Gymnastics: State that just as the human body can change shape and orientation, the rotational inertia of the body changes also. Discuss the figures in text Section 11.5 and how rotational inertia is different for the same body configuration about different axes, as illustrated in Figure 11-20. Discussion of this section will be of considerable interest to your students, because it's about them and their interests in a direct way.

Angular Momentum and Its Conservation: Just as inertia and rotational inertia differ by a radial distance and just as force and torque also differ by a radial distance, so do momentum and angular momentum differ by a radial distance.

Relate linear momentum to angular momentum for the case of a small mass at a relatively large radial distance — the object you previously swung overhead.

For the more general case, angular momentum is simply the product of rotational inertia I and angular velocity ω. This is indicated in text Figure 11-27.

> DEMONSTRATION [11-10]: With a weight in each hand, rotate on a platform as shown in Figure 11-27. Show angular momentum conservation by drawing in your arms and thus speeding up.

Cite the fact that much mass flows down the Mississippi River as mud that is deposited in the Gulf of Mexico. Ask what effect this tends to have on the earth's rotation. [It tends to slow the earth, which you can simulate on the rotating table with one weight held outstretched at about 45° and then lowered to horizontal position. Students will see a slowing of your rotational speed.] Cite also the fact that when polar ice melts, the melted water tends to flow toward the earth's equator, thus effectively spreading the mass of ice away from the polar axis. Ask what effect this tends to have on the earth's rotation. [Again, it tends to slow the earth, which you can simulate on the rotating table with a weight in each hand held outstretched over your head and then lowered to a horizontal position. Students will see even more slowing of your rotational speed.]

> DEMONSTRATION [11-11]: Show the operation of a gyroscope — either with a model or a rotating bicycle wheel as is demonstrated in text Figure 11-24. You can demonstrate angular momentum conservation nicely if you stand on the turntable and show different orientations of the spinning wheel. Begin with the axes of a spinning bike wheel and stationary rotating platform perpendicular [no effect]. Then turn the axis of the spinning bike wheel so it is parallel to the axis of the platform. If you had no angular momentum initially, you'll rotate in a direction opposite to that of the spinning wheel to produce the

same zero angular momentum. Different angles produce different components of angular momentum with interesting effects. The angular momentum vector is along the axis of rotation. Let the spin be in the direction of your curled fingers on your right hand. Then the angular momentum vector is in the direction of your thumb.

More Think-and-Explain Questions

1. If you place any weight on a balanced seesaw, it will topple unless the weight is placed directly above the fulcrum. Why will weight added above the fulcrum not upset the balance?

   Answer: At this place, the added force passes through the CG of the seesaw and the fulcrum; no lever arm — no torque — no rotation.

2. A basketball player wishes to balance a ball on his fingertip. Will he be more successful with a spinning ball or a stationary ball? What physical principle supports your answer?

   Answer: A spinning ball is more easily balanced, in accordance with the law of rotational inertia: A spinning body tends to remain spinning.

3. Suppose you sit in the middle of a large freely-rotating turntable at an amusement park. If you crawl toward the outer rim, does the rotational speed increase, decrease, or remain unchanged? What physical principle supports your answer?

   Answer: Rotational speed decreases, in accordance with the conservation of angular momentum.

4. Why is it incorrect to say that when you execute a somersault and pull your arms and legs inward, your angular momentum increases?

   Answer: It is your rotational speed that increases, not your angular momentum. Angular momentum remains the same. In this chapter, confusion is more likely to occur with the many terms, than with the concepts.

To use this planning guide work from left to right and top to bottom.

<table>
<tr><td colspan="4" align="center">**Chapter 12 Planning Guide**
• *The bulleted items are key: Be sure to do them!*</td></tr>
<tr><td>**Topic**</td><td>**Exploration**</td><td>**Concept Development**</td><td>**Application**</td></tr>
<tr><td>**Falling Apple, Moon, Earth**</td><td></td><td>Text 12.1–12.3/Lecture</td><td></td></tr>
<tr><td>**Gravity Formula**</td><td></td><td>• Text 12.4/Lecture</td><td></td></tr>
<tr><td>**Inverse-Square Law**</td><td></td><td>• Text 12.5/Lecture
Con Dev Pract Pg 12-1</td><td>Nx-Time Q 12-1</td></tr>
<tr><td>**Universal Gravitation**</td><td></td><td>• Text 12.6/Lecture</td><td></td></tr>
<tr><td colspan="4">**Video:** *Gravity*
Evaluation: Chapter 12 Test</td></tr>
</table>

See the Chapter Notes for alternative ways to use all these resources.

Objectives

After studying Chapter 12, students will be able to:
• Explain Newton's idea that the moon falls toward the earth like an apple does.
• Explain why the moon does not fall into the earth and the planets do not fall into the sun.
• State Newton's law of universal gravitation.
• Explain the significance of an inverse-square law.
• Explain the connection between gravitation and the idea that the universe may stop expanding and begin to contract.
• Give examples of how Newton's theory of gravitation affected the thinking of philosophers of the eighteenth century.

Possible Misconceptions to Correct

• Newton discovered gravity rather than discovering that gravity was universal.
• Above the atmosphere of the earth, there is no earth gravity.
• The moon and planets are beyond the pull of the earth's gravity.

No Demonstrations for this Chapter

Introduction

I recommend the delightful book *The Attractive Universe*, by E.G. Valens and Berenice Abbott (World, 1969). Good material!

For a more thorough treatment of the falling apple and the falling moon, text pages 162-165, see Volume I of *The Feynman Lectures on Physics*, pp. 7-9. The fact that an object on earth released from rest falls a vertical distance of 4.9 m in the first second (Chapter 2) is utilized here in the development of Newton's theory of gravitation. This chapter presents a good climate for discussing the meaning of a scientific theory, (see the footnote on page 137). This is expanded nicely in the last chapter on *Cargo Cult Science* of Feynman's book *Surely You're Joking, Mr. Feynman* (Norton, 1985). Along with this entertaining book, read also Feynman's *What Do You Care What Other People Think* (Norton, 1987).

Although the formula for Newton's law of gravitation is not shown until five pages into the chapter, I have found considerable success in lecture by beginning with the law right away. All the examples throughout the lecture relate to the formula. The formula focuses on what might be seen as diverse phenomena. (Acknowledge that many other texts and references use the symbol r instead of the d used in this text. The r is used to indicate the radial distance from a body's CG and to emphasize the center-to-center, rather than the surface-to-surface, nature for distance, and to prepare for r used as a displacement vector. We don't set our plow that deep, however, and we use d for distance.)

There are no demonstrations for this chapter. Keep class participation and interest via the check-your-neighbor routine. (Have you noticed by now that this routine allows you ample time to check your own notes, collect your thoughts, and improve your timing?)

Suggested Lecture

Begin by briefly discussing the simple codes and patterns that underlie the complex things around us, whether they be musical compositions or DNA molecules. Then briefly describe the harmonious motion of the solar system and the Milky Way and other galaxies in the universe by stating that the shapes of the planets, stars, and galaxies, and their motions are all governed by an extremely simple code or pattern. Next write the formula for universal gravitation on the board.

Formula Supported by Examples: Give examples of bodies pulling on each other to convey a clear idea of what the symbols in the equation mean and how they relate.

CHECK QUESTIONS: How is the gravitational force between a pair of planets altered when the mass of one is doubled? [Twice the force] When both are doubled? [The force is four times as great.] When they are twice as far apart? [When twice as far apart, the force decreases to 1/4 as much.] When they are three times as far apart? [Force is 1/9 as much.] Ten times as far apart? [Force is 1/100 as much.]

Development of the Formula: Discuss how Newton developed the law by going from falling apples to the falling moon. Explain what is meant by *tangential* speed or velocity. The physics of the falling earth is explained in more detail in Chapter 14. (Call attention to the comic strip, "Satellite Physics" on text page 192 if questions are raised about satellite motion.)

Inverse-Square Law: Discuss the inverse-square law and go over text Figure 12-10, or its equivalent, for candlelight or radioactivity.

CHECK QUESTIONS: A space probe is a certain distance, center to center, from a massive star. If it is four times as far from the star, how will its gravitational force toward the star compare? [It is 1/16 as much.] A sheet of photographic film is exposed to a point source of light that is a certain distance away. If the sheet was exposed to the same light four times as far away, how would the intensity on the film compare to its original exposure? [It would be 1/16 as much.] A radioactive detector registers a certain amount of radioactivity when it is a certain distance away from a small piece of uranium. If the detector is four times as far from the uranium, how will the radioactivity reading compare to the original? [It will be 1/16 as much.]

On the board, plot an inverse-square curve to scale. Show the steepness of the curve as it goes "suddenly" from 1/4 to 1/9, to 1/16 for a twice, thrice, and four times separation. This is shown in text Figure 12-11.

CHECK QUESTIONS: True or false. The force of the earth's gravity on the space shuttle in orbit is zero, or nearly zero. [False! The force of the earth's gravity on the shuttle in orbit is nearly the same as the force of earth's gravity on the shuttle at sea level. At an altitude of 200 km, well above the earth's atmosphere, the space shuttle is only 3% farther from the earth's center and experiences 94% the gravitational pull at the earth's surface.] True or false. At the far reaches of the universe, a body will experience zero earth gravity. [False! The equation guides thinking here. As distance d approaches infinity, force F approaches (but does not reach) zero. As a practical matter, at such a distance, the force due to the earth's gravity may be negligible in comparison to the influences of closer and more massive bodies. Strictly speaking, however, the gravitation of the earth extends to infinity. No matter how far you go, earth gravity is your companion.]

Gravitational Constant G: Explain the gravitational constant G by comparing it to π, the constant for circles. Begin by writing $C \sim D$, draw several different-size circles on the board and show how circumference and diameter are proportional. State that if you divide the circumference C by the diameter D for any circle, you get the same number, 22/7. This constant number is called π. So the proportion can be written as the exact equation, $C = \pi D$. This is similar for the constant G in Newton's equation. When the force of gravity, F, between two bodies of mass, m_1 and m_2, separated by distance d is divided by $m_1 m_2/d^2$, the number that results is a constant (6.67 x 10^{-11} Nm²/kg²). So the proportion $F \sim m_1 m_2/d^2$ can be written as the exact equation $F = G m_1 m_2/d^2$. Call attention to von Jolly's method in text Figure 12-8.

Universal Gravitation: Discuss the theory of expanding matter in the universe and its possible oscillating mode. You can get class interest into high gear with speculations about the possibility of past and future cycles. After discussion, you can end your class on a high note by doing the following. State that you wish to represent a single cycle with the positions of the chalk you are holding in your hand above the lecture table. Place the chalk on the table and let it represent the time of the primordial explosion. Then, as if the chalk were projected upward, raise the chalk to a point just above your head and state that this position represents the point where the universe momentarily stops before beginning its inward collapse, the "big crunch." Then move your hand slowly down, speeding up and back to its starting point to indicate the completion of one cycle. Then hold the chalk a foot or so above the table to show a point corresponding to our present location — a point representing about 15 to 20 billion years from the beginning of the cycle. Holding the chalk steady and purposefully, ask where the chalk should be positioned to represent the dawning of civilization. Then move the chalk to a position about a quarter of an inch below the present point, stating that's where we were and this is where we are, as you move the chalk back to the present position. Still holding it there, ask students to speculate on where humankind will be and what the world will be like when we move another quarter of an inch as you move the chalk upward to show that position. That point, of course, represents a time on earth difficult to comprehend.

More Think-and-Explain Questions

1. The earth and the moon gravitationally attract each other. Does the more massive earth attract the moon with a greater force, the same force, or less force than the moon attracts the earth?

   Answer: The force on both is the same, in accordance with Newton's third law. Note also from Newton's formula for gravitation that the force does not depend on the order of the masses, but depends simply on the product.

2. What is the magnitude and direction of the gravitational force that acts on a woman who weighs 500 N at the surface of the earth?

   Answer: 500 N straight downward: her weight

Computational Problems

1. A light source located 1 m away from an opening which is 1 square meter will cover 4 m² when it is 2 m away from the opening.

How many square meters will it cover when it is 3 m away? 10 m away?

Answers: At 3 m it will cover 3², or 9 square meters; at 10 m, it will cover 10², or 100 square meters.

2. The value g at the earth's surface is 9.8 m/s². What is the value of g at a distance from the earth's center that is 4 times the earth's radius?

   Answer: $g/16$, or 0.6 m/s²

3. Calculate the force of gravity between the earth (mass = 6 x 10^{24} kg) and the sun (mass = 2 x 10^{30} kg, distance = 1.5 x 10^{11} m).

   Answer: $F = GmM/d^2$ = (6.67 x 10^{-11} Nm²/kg²) (6 x 10^{24} kg)(2 x 10^{30}kg)/(1.5 x 10^{11}m)² = 3.6 x 10^{22} N.

4. Suppose the force of gravity between the sun and earth vanishes. Instead, a steel cable between these two bodies keeps the earth in orbit. Estimate the thickness (diameter) of such a cable. You need to know that the strength of steel is 2 x 10^{11} N/m² (called the elastic modulus). That is, a steel cable with a cross-sectional area of 1 m² will support a force of 2 x 10^{11} N.

   Answer: 480 km thick! From the ratio 3.6 x 10^{22} N/x = 2 x 10^{11} N/1 m², x = (3.6 x 10^{22})/(2 x 10^{11}) = 1.8 x 10^{11} m². This would be the cross-sectional area of the cable. From the area of a circle, $A = \pi D^2/4$, we find its diameter $D = \sqrt{4A/\pi}$ = 4.8 x 10^5 m = 480 km.

5. With how much force is a 1-kg mass on earth attracted to the moon? (The moon's mass is 7.4 x 10^{22} kg and its distance is 3.8 x 10^5 km.)

   Answer: $F = GmM/d^2$ = (6.67 x 10^{-11})(1)(7.4 x 10^{22})/(3.8 x 10^8)2 = 3.4 x 10^{-5} N.

6. Use the results of the previous problem to calculate how great a mass is if, when located 1 m away from the 1-kg mass, it exerts as much gravitational force on it as the moon does.

   Answer: About 5 x 10^5 kg. From $F = GmM/d^2$, where m = 1 kg, d = 1 m, and F = 3.4 x 10^{-5} N, $M = Fd^2/Gm$ = (3.4 x 10^{-5})(1)²/[(6.67 x 10^{-11})(1)] = 5 x 10^5 kg. (Such a mass weighs about 500 tons! Practically speaking, it would be quite incredible to arrange a distance of separation of only 1 m for so massive an object.)

7. If you stand 1 kilometer from the base of a typical mountain with a mass of 5 x 10^{11} kg, you'll be gravitationally attracted to it. Likewise, you're attracted to the moon (7.4 x 10^{22} kg of mass and 3.8 x 10^5 km distant). Which of these two exerts the greatest force on you?

   Answer: About the same. Consider the ratio of mountain F to moon F': F/F' = (GmM/d^2) $/(GmM'/D^2)$ = $(M/d^2)/(M'/D^2)$ = [(5 x10^{11}) /(1)²]/[(7.4 x 10^{22})/(3.8 x 10^5)²] = (5 x 10^{11})/(5 x 10^{11}) = 1. (Note here that because they cancelled anyway, we left the distances in km. Their ratio would be the same in meters or any other pair of consistent units. Note also that since only the ratio is considered, no units are shown here.)

13 Gravitational Interactions

To use this planning guide work from left to right and top to bottom.

Chapter 13 Planning Guide
• *The bulleted items are key: Be sure to do them!*

Topic	Exploration	Concept Development	Application
Gravitational Fields		• Text 13.1–13.2/Lecture Con Dev Pract Pg 13-1	Exp 36 (1period) Exp 37 (1period)
Weightlessness	• Act 38 (1 period)	• Text 1.3/Lecture Con Dev Pract Pg 13-2	
Tides		Text 13.4–13.5/Lecture	
Black Holes		Text 13.6/Lecture	Nx-Time Q 13-1
Video: *Gravity 2* **Evaluation:** Chapter 13 Test			

See the Chapter Notes for alternative ways to use all these resources.

Objectives

After studying Chapter 13, students will be able to:
• Distinguish between **g** (the acceleration due to gravity) and *G* (the universal gravitational constant).
• Describe a gravitational field.
• Describe the gravitational field of the earth both inside and outside the earth's surface.
• Explain why an astronaut in earth orbit seems weightless even though there is a gravitational force on the astronaut.
• Explain how the moon and sun cause the ocean tides.
• Give examples of tides other than those in water.
• Describe how a black hole is formed.

Possible Misconceptions to Correct

• The fact that the same side of the moon faces earth is evidence that the moon doesn't spin about its axis.
• The crescent shape of the moon is created by the earth's shadow.
• The fact that the moon is the chief cause of ocean tides is evidence that the moon's pull on the earth is greater than the sun's pull is.

No Demonstrations For this Chapter

Introduction

The concept of force field introduced in this chapter is a good background for the electric field treated later in Chapter 33. The gravity field here is applied to regions outside as well as inside the earth. You may expand on the "tunnel all-the-way-through-the-earth" bit and explain how, ideally, the period of oscillation of a body traveling in such a tunnel under the influence of only gravity would be the same for any straight tunnel — whether it's from New York to Hawaii or from New York to China. You can support this with the analogy of a pendulum that swings through different amplitudes with the same period. In nonvertical tunnels, of course, the object must slide rather than drop without friction. The timetables for travel in this way would be quite simple; any one-way trip would take 43 minutes!

The answer to Check Question 1 on text page 178 states without explanation that a body dropped in a tunnel bored through the earth will undergo simple harmonic motion. The condition for simple harmonic motion is that the restoring force be directly proportional to displacement, as is the case for a bob at the end of a spring (restoring force is $-kx$, Hooke's law). It turns out that the force of gravity on a body inside a planet of uniform density is directly proportional to the distance from its center (not the distance squared from its center). Here's the explanation: We know that the gravitational force F between a particle m and a spherical mass M, when m is outside M, is simply $F = GmM/d^2$. But when m is inside a uniformly-dense solid sphere of mass M, the force on m is due only to the mass M' contained within the sphere of radius r ($< R$), represented by the dashed line in the figure. Contributions from the shell $>r$ cancel out. So, $F = GmM'/d^2$. From the ratio of M'/M, you can see that $M' = Mr^3/R^3$, [that is, $M'/M = V'/V = (4/3 \pi r^3)/(4/3 \pi R^3) = r^3/R^3$]. Substitute M' in Newton's equation for gravitation and you get $F = GmMr/R^3$. All terms on the right are constant except r. So $F = kr$; force is linearly proportional to radial distance when $r < R$.

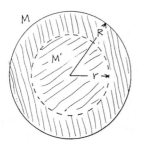

Compared to the gravitational attraction of local buildings and the like, the pull to the moon is appreciable. Consider the ratio of the mass of the moon to its distance squared:

$$7.4 \times 10^{22} \text{kg}/(4 \times 10^5 \text{km})^2 = 5 \times 10^{11} \text{kg/km}^2$$

This is a sizable ratio, one that buildings in your vicinity cannot match (city buildings of greatest mass are typically on the order of 10^6 or 10^7 kilograms). However, if you stand 1 kilometer away from the foot of a mountain of typical mass 5×10^{11} kilograms, then the pull between you and the mountain and the pull between you and the moon are the same. Interestingly enough, with no friction you would tend to gravitate from your spot toward the mountain — but you experience no tendency to gravitate from your spot toward the moon! That's because the spot you stand on undergoes the same gravitational acceleration toward the moon as you do. Both you and the whole earth are in lunar free fall, accelerating toward the moon. Whatever the

lunar force on you, it has no tendency to pull you off a weighing scale — which is why your weight has nothing to do with the positions of the sun or moon. When you step on a weighing scale, the interaction is only between you and the mass of the earth.

It is interesting to expand upon the zero-g field inside a hollow planet and to speculate about the living conditions of a civilization in such an environment. If the shell is uniform, g is zero everywhere inside and not just at the center as indicated in text Figure 13-5.

This chapter is a good place for students to learn some of the principles of creating a scientific theory by means of an imaginary model — for example, ocean tides. The main points are (1) make the model simple, (2) consider only one effective element at a time, (3) explore the limits of the variables from small to large (or from zero to infinity), and (4) take time, proceed step by step, and give imagination a chance.

A brief treatment of black holes is included in this chapter. It is interesting to note that light bends in any gravitational field and not simply in the enormous gravitational fields near black holes. Einstein stated that light bends in a gravitational field just as a thrown baseball does. According to Einstein, if you could throw a baseball as fast as light (you can't), both would follow the same trajectory! We say that light travels in straight lines for much the same reason some people say that a high-speed bullet doesn't curve downward in the first part of its trajectory. Over short distances, the bullet doesn't *appear* to drop only because of the short time involved. The same is true with light, but we don't notice because of the vast distance involved compared to the brief time it is in the strong part of the earth's gravitational field.

While the information in this chapter is useful as a background for force fields in general, it may be skipped without complicating the treatment of other material. This is an interesting chapter, because the material is interesting by itself. Also, it is interesting historically and is closely related to space science, which is currently in the public eye.

Suggested Lecture

G, g, and $\mathbf{g}$:   Distinguish between the universal gravitational constant G, the acceleration due to gravity g, and the gravitational field vector $\mathbf{g}$. Although $\mathbf{g}$ and g represent different concepts, they have the same numerical value, the familiar 9.8 m/s^2. It is not important to emphasize the distinction between $\mathbf{g}$ and g since they can be applied interchangeably. It is very important however, to stress the distinction between G and $\mathbf{g}$ (or g). G and $\mathbf{g}$ represent completely different quantities. Their relationship is presented on text page 176, where G is derived from $\mathbf{g}$. This is the first derivation in the text not tucked away in a footnote.

Relate the derivation of G to Philipp von Jolly's method of finding G, shown in text Figure 12-8. Sketch Figure 12-8 on the board and explain that all values needed to calculate the mass of the earth except G were measured before von Jolly's experiment. Explain how once G was found, the mass of the earth was known also. (Equate the weight of anything, mg, to the force of gravity, GmM/d^2. Then $M = gd^2/G$, where d is the earth's radius, g is 9.8 m/s², and G is 6.67×10^{-11} Nm²/kg².)

CHECK QUESTION: If the earth had its same size and twice its mass, what would be the acceleration of freely-falling objects at its surface? [Twice g, or nearly 20 m/s²] If the earth had its same mass and half its size, what would be the acceleration of freely-falling objects at its surface? [Four times g, or nearly 40 m/s²] If the earth had twice its mass and half its size? [Eight times g, or nearly 80 m/s².]

Gravitational Field: Show or call attention to the altered space that surrounds a magnet — a kind of aura called a magnetic field. A magnetic field is a force field, because magnetic materials in it experience a force. The same is true with the gravitational field about the earth or any mass. A mass in the field region experiences a gravitational force. The force field is strongest at the surface of the earth, and its strength is less as the inverse-square of distance from the earth's center.

CHECK QUESTIONS: What evidence would you look for in order to determine whether or not you were in a gravitational field? [The presence of a gravitational force] How strong is the gravitational field twice as far from the earth's center when compared to the gravitational field at the surface of the earth? [By the inverse-square law, 1/4 as strong]

Field Inside the Earth: It is interesting to consider a tunnel bored clear through the earth. It's easy to convince your students that the gravitational force on a body located at the exact center of the tunnel is zero. A chalkboard sketch showing a few symmetrical force vectors does this. The gravitational field at the earth's center is zero. Consider the magnitude of force the body would experience away from the center, somewhere between the center and the surface. A few more carefully-drawn vectors will show that the forces don't cancel to zero. The gravitational field is between zero and its value at the surface. If the density of the earth were uniform, the field would be linear from the center to the surface. The earth's field would be half at the halfway mark, 3/4 at the 3/4 mark and so on, as shown by the straight line in the sketch. (Actually the field is

not linear, because the density is much greater at the earth's center. The shape of the curve is actually as indicated in the right-hand sketch.)

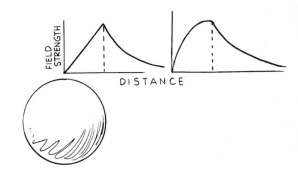

Earth Tunnel: You may want to go beyond the text coverage and discuss the motion of a body dropped in a tunnel bored completely through the earth and how it would keep rhythm with a circularly-moving satellite of the same "amplitude." It takes nearly 90 minutes for a satellite to make a complete trip around the earth in a close orbit, which is the exact amount time it would take a body dropped from the same altitude into the tunnel to go through the earth (see text page 177). It should be enough to state the oscillating case without going into the mathematics of simple harmonic motion, which is discussed briefly in the second paragraph at the beginning of this *Teaching Guide* chapter.

CHECK QUESTION: If you dropped a rock into a tunnel through the earth, would the rock shoot out the other side like it was fired from a "gravity gun"? [No, it would gain speed until it reached the earth's center, and then it would lose speed the rest of the way. Its speed at the far end of the tunnel would be the same as its initial speed. It would then fall back and repeat the motion in cyclic fashion.]

CHECK QUESTION: As the velocity of the falling body in the earth tunnel increases, does the acceleration increase, decrease, or remain unchanged? [The acceleration decreases as the gravitational field decreases and is zero at the earth's center. The falling body has its maximum velocity at the earth's center, where both the field and acceleration are zero.]

Weight and Weightlessness: Define weight in terms of support force. According to this definition, we are as heavy as we feel. Contrast this to apparent weightlessness, and relate it to the queasy feeling your students experience when in a car that goes too fast over the top of a hill. State that this feeling is what an astronaut in orbit is confronted with all the time! Ask how many in your class would still welcome the opportunity to take a field trip to Cape Canaveral and take a ride aboard the shuttle. What an exciting prospect!

CHECK QUESTION: Why would you feel weightless in an elevator with a broken cable? [There would be an absence of a support force — the floor would fall as fast as you do.]

Freely-Falling Elevator: Cite the apparent weightlessness of astronauts orbiting in the space shuttle and how videocasts show things floating around as if no gravity was present. Imitate an astronaut removing a pen from his or her shirt pocket and releasing it, only to find it floating where it is released. Then ask your class to consider a video camera fixed to the inside of an elevator. Pretend you are in the elevator and remove your pen and drop it. The video records the dropping of the pen. No big deal. But now consider what the camera would see if you repeated the pen-drop maneuver in an elevator that is in free fall. The camera would show the pen floating beside you as you, pen, camera, elevator, and all fall at g. Ask if there is a force of gravity on you in this case, as evidenced by the sudden stop! Does the camera show the dropping motion? Isn't this what occurs in orbit? Viola!

(This is a good place to break.)

Ocean Tides: Begin your treatment of tides by asking the class to consider the consequences of someone pulling at your coat. If they pulled only on the sleeve, for example, it would tear. But if every part of your coat were pulled equally, it and you would accelerate but it wouldn't tear. It tears when one part is pulled harder than another, or it tears because of a *difference* in forces acting on the coat. In a similar way, the spherical earth is "torn" into an elliptical shape by the differences in gravitational forces by the moon. The moon's gravitational forces are stronger between the moon and the near side of the earth, and weaker between the moon and the far side of the earth. A similar situation exists for the sun, but the differences in force are less.

Explain that the moon "out-tides" the stronger-pulling sun, because the difference in pulls on either side of the earth is greater for the closer moon. Explain that tides are extra-high when the moon and sun are lined up, because the pulls add and the two tides caused by the moon and sun overlap. Explain that tides are not as high when the moon and sun are at right angles to each other, because the high tide of the sun overlaps the low tide of the moon, and vice versa.

CHECK QUESTION: Which pulls harder on the oceans of the earth, the sun or the moon? [The sun] Which is most effective in raising tides, the sun or the moon? [The moon] (This question is a good one for determining who is reading the book and who isn't!)

Explain how the earth turns daily beneath the two ocean bulges to give us two high and two low tides per day. This is best discussed with a globe of the earth on your lecture table. Designate an object across the room as the moon and a high tide on the earth in its direction. Then rotate the globe and show how different parts of the earth pass "beneath" the moon and experience high tides. A similar situation occurs on the opposite side of the earth.

CHECK QUESTION: At the time of extra-high tides, will extra-low tides follow in the same day? [Yes, by the "conservation of water"; There is only so much water on the earth, so extra-high tides in one part of the world mean extra-low tides in another. If students don't see this, ask them to imagine that are sloshing water back and forth while taking a bath in a bathtub. When the water is extra deep in the front part of the tub, doesn't this mean the water level will be extra low in the back part? The same is true for the ocean!]

Misconceptions About the Moon: This is an appropriate place for you to dispel two popular misconceptions about the moon: (1) Since one side of the moon's face is "frozen" to the earth, it doesn't rotate about its polar axis. (2) The crescent shape commonly seen is the earth's shadow. To convince your class that the moon does rotate about its polar axis, simulate the situation by holding your eraser at arms length in front of your face. Tell your class that the eraser represents the moon and your head represents the earth. Rotate slowly keeping one face of the eraser in your view. Call attention to the fact that from your frame of reference, the eraser doesn't rotate as it revolves about you — as evidenced by your observation of only one face, with the backside hidden. However, your students occupy the frame of reference of the stars (each of them *is* a star). From their point of view, they can see all sides of the eraser as it rotates, because it turns about its axis as often as it revolves about yours. Show them how the eraser, if not slowly rotating (frozen with one face always facing the same stars), would show all of its sides to you as it circles around you. See one face, then wait 14 days later and the backside is in your view. The moon has a rotation rate that is the same as its revolution rate. The footnote on text page 182 gives an explanation for the same side of the moon facing earth.

To correct the second misconception, draw a half-moon on the board. The shadow is along the diameter and is perfectly straight. If that were the shadow of the earth, then the earth would have to be flat or a big block shape! Discuss playing "flashlight tag" with a suspended basketball in a dark room that is illuminated by a flashlight in various locations. Ask your class if they could estimate the location of the flashlight by only looking at the illumination of the ball. The same is true with the moon illuminated by the sun!

CHECK QUESTION: On the board sketch the picture below and ask what is wrong with it. [The moon is in a daytime position, as evidenced by the upper part of the moon being illuminated. This means the sun must be above the horizon (at about "11 o'clock"). Dispel the notion that the crescent shape of the moon is a partial eclipse by considering a half-moon and the shape of the earth necessary to cast such a shadow.]

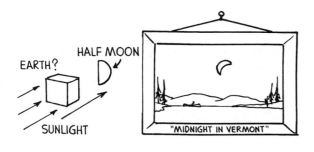

Suggest that students observe the phase of the moon and compare it to the phases shown on wall calendars.

Tides in the Earth and in the Atmosphere:

For the same reason that we have ocean tides, there are tides of the earth, which, after all, is mostly molten lava. Mount Everest rises and falls about 30 cm twice a day. Interestingly enough, there is a greater probability for earthquakes and volcanoes when there is an eclipse of the sun or moon. This is most probable when the earth experiences neap tides and greater stresses are on the earth's crust.

After your treatment of tides as a result of the sun and moon pulling on the earth, address yourself to text Think and Explain Question 9 concerning biological tides. Many people attribute special significance to the gravitational pull of the moon on themselves, and they support this with the fact that the moon raises the ocean an average of 1 meter each 12 hours. Repeat the reason for the tides; half the water is closer to the moon than the other half is. In order to be subject to any measurable lunar tides in our bodies, part of our bodies must be appreciably closer to the moon than other parts are. Detectable lunar tides in body fluids would occur only in tall people — like many kilometers tall!

Add to the text material by discussing the consequences if the moon were closer. Ocean tides would be higher, and by the same token, the tidal forces on the moon's crust would be greater. If they were too close, the earth's tidal forces would tear the moon into a billion pieces, forming a ring around the earth similar to that of Saturn. Saturn's rings are thought to be the debris left over from bodies torn apart by tidal forces.

Black Holes: Tidal forces reach an extreme in the case of a black hole. The unfortunate fate of an astronaut falling into a black hole is not encountering the singularity, but the tidal forces encountered long before getting that close. Approaching feet first, for example, the astronaut's feet, being closer, would be pulled with a greater force than his midsection, which in turn would be pulled with a greater force than his head. The tidal forces would stretch the astronaut who would be killed before these forces literally pulled his body apart.

Describe how a collapsed star represents condensed mass and therefore condensed gravity. The mass of a black hole is no greater than the mass of the star which collapsed to form it. Hence the gravitational field of the star and the black hole are the same at distances greater than the original radius of the star. It is only at closer distances that the enormous field occurs. Discuss Figure 13-19 on text page 187.

CHECK QUESTION: Consider a satellite that is companion to a star which collapses to become a black hole. How will the orbit of the companion satellite be affected by the star's transformation to a black hole? [It's not affected at all, because no terms in the gravitational equation change.]

More Think-and-Explain Questions

1. The gravitational field strength at the earth's surface is 9.8 N/kg. What is the gravitational field strength at the center of the earth? At a distance of 1 earth-radius beyond the surface?

   Answers: Zero at the earth's center, and 1/4 or 2.45 N/kg when twice as far from the center, or 1 earth-radius from the surface

2. How would the gravitational field at the earth's surface be affected if the earth shrank in size without any change in its mass? What would be its relative strength at the new surface if the earth shrank to half its size? To one-tenth its size?

   Answers: At half its size, strength would be increased by 4; at one-tenth, strength would be 100 fold. (Can you see why gravity is so intense at the surface of a collapsed star?)

3. How is stepping off a curbstone similar to taking a ride aboard the space shuttle?

   Answer: In both cases you experience weightlessness. The principle difference is in the time involved. Stepping off a curbstone involves an apparent weightlessness that is so brief, it is not noticed.

4. The sun exerts almost 200 times more force on the oceans of the earth than the moon does. Why then, is the moon more effective in raising tides?

   Answer: Tides are caused by the *difference* in gravitational pulls. The moon pulls with proportionally more force on the near side of the earth to the moon than on the far side. This difference in the pulls is greater than the corresponding difference in the pulls by the more distant sun.

5. From a point of view at the sun, does the moon circle the earth, or does the earth circle the moon?

   Answer: Neither; both the earth and the moon circle a common point, the center of mass of the earth-moon system (called the barycenter), which is located about three quarters of the earth's radius from its center at about 1600 km below the earth's surface. It is the center of mass that smoothly orbits the sun, while the earth and moon wobble monthly about this center.

6. What would be the effect on the earth's tides if the diameter of the earth was very much larger than it is? What would be the effect if the earth was as it presently is and the moon was very much larger, but with the same mass?

   Answers: Tides would be greater if the earth's diameter were greater, because the difference in the gravitational pulls would be greater. Tides on earth would be no different if the moon's diameter was larger. The gravitational influence of the moon is as if all the moon's mass were at its CG. Tidal bulges on the solid surface of the moon, however, would be greater if the moon's diameter was larger — but not the tidal bulges on the earth.

Computational Problems

1. The mass of Saturn is 95 times that of the earth, and its radius is 9 times that of the earth. Calculate the acceleration due to gravity at the surface of Saturn. Express your answer in g's.

   Answer: About 1.2 g. On earth, $g = GM/R^2$, where M is the earth's mass, and R its radius. Saturn's mass is $95M$, and its radius is $9R$. Then on Saturn, the acceleration due to gravity $= G(95M)/(9R)^2 = 95/81\ GM/R^2 = 95/81\ g = 1.2\ g$.

2. The mass of the earth is about 80 times that of the moon and its radius is about 3.7 times that of the moon. Calculate the acceleration due to gravity at the surface of the moon. Express your answer in g's.

   Answer: About 0.17 g. On earth, $g = GM/R^2$, where M is the earth's mass, and R its radius. Moon's mass is $M/80$, and its radius is $R/3.7$. Then on the moon, the acceleration due to gravity $= G(M/80)/(R/3.7)^2 = 3.7^2/80\ GM/R^2 = 13.7/80\ g = 0.17\ g$. (That's about $g/6$.)

3. If you drop a 1-kg mass just above the earth's surface, it accelerates downward at 9.8 m/s². But the force that pulls the 1-kg mass downward is also the force that must pull the 6 x 10²⁴-kg earth upward. Compute the acceleration of the earth as it "races upward" to meet the 1-kg mass.

   Answer: $$a = \frac{F}{M} = \frac{mg}{M}$$

   $$= \frac{1\ \text{kg} \times g}{\left(6.2 \times 10^{24}\ \text{kg}\right)} = \frac{9.8\,\text{N}}{\left(6.2 \times 10^{24}\ \text{kg}\right)}$$

   $$= 1.6 \times 10^{-23}\ \text{N} / \text{kg}$$

   $$= 1.6 \times 10^{-23}\ \text{m} / \text{s}^2 \left(\text{since N} / \text{kg} = \text{m} / \text{s}^2\right)$$

That's why the upward acceleration of the earth is not observable — it's much too small!

To use this planning guide work from left to right and top to bottom.

Chapter 14 Planning Guide
• *The bulleted items are key: Be sure to do them!*

Topic	Exploration	Concept Development	Application
Earth Satellites		• Text 14.1/Lecture	
Circular Orbits		• Text 14.2/Lecture	Nx-Time Q 14-1
Elliptical Orbits	Act 39 (1 period)	• Text 14.3/Lecture Con Dev Pract Pg 14-1 • Exp 40 (1 period)	
Energy Conservation		Text 14.4/Lecture	Nx-Time Q 14-2
Escape Speed		Text 14.5/Lecture	
Video: *Satellite Motion* **Evaluation:** Chapter 14 Test			

See the Chapter Notes for alternative ways to use all these resources.

Objectives

After studying Chapter 14, students will be able to:
• Explain how the speed of a satellite in a circular orbit around the earth is related to the distance an object falls in the first second due to gravity.
• Explain why the force of gravity does not cause a change in the speed of a satellite in circular orbit.
• Describe how the speed of a satellite changes for different portions of an elliptical orbit.
• Apply the energy conservation law to describe changes in the potential and kinetic energies of a satellite for different portions of an elliptical orbit.
• Describe what is meant by an escape speed.

Possible Misconceptions to Correct
• Satellites are beyond the main pull of a planet's gravitational field.
• Satellites are held up by a centrifugal force.

Demonstration Equipment
• [14-1] Piece of string with two suction cups or tacks to show the construction of an ellipse

Introduction

The idea of satellite motion as an extension of projectile motion was introduced in Chapter 6, and the idea of the "falling moon" was introduced in Chapter 12. The comic strip, *Satellite Physics* on text page 192 says it all. The rest is embellishment.

The excellent NASA 15-minute film, *Zero-g*, is a must. It is historic footage taken aboard Skylab in 1978 and narrated by astronaut Owen Garriott.

An overview of Newton's laws of motion is treated with excellent examples and with a touch of humor. Your students will enjoy this educational film.

Note that Kepler's laws are not covered in the text. Your students will "discover" Kepler's third law by using the computer and the lab *Trial and Error*. It is best to introduce Kepler and his laws of planetary motion after your students do this great computer activity.

When discussing the elliptical paths of satellites, you may point out that when we toss a baseball into the air we say its path is parabolic, but strictly speaking, the path is a segment of an ellipse. The earth's center is at the far focus of this ellipse. If nothing were in the way, the baseball would follow an eccentric elliptical path and return to its starting point! The earth's center is at the near focus for satellites that trace external elliptical paths around the earth. In this case, nothing occupies the other focal point.

Suggested Lecture

Introduce satellite motion as an extension of simple projectile motion, as shown in the comic strip "Satellite Physics." On the board draw a world with a hypothetical mountain at the top, as shown in the sketch. Call this "Newton's Mountain," and show it high enough to poke through the atmosphere so that cannonballs fired from it encounter no air resistance. Show how successively greater speeds result in a circular orbit. State that Isaac Newton thought of this idea and calculated the required speed of a cannonball for circular orbit. State that you expect that many students in your class will be able to do the same before the class period is over. You will provide them with some information about the world that will help them make the calculation.

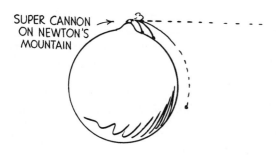

SUPER CANNON ON NEWTON'S MOUNTAIN

Calculation of Orbital Speed: Sketch text Figure 14-4 on the board. Pretend to place a laser on a 1-meter-high tripod and to aim it over a perfectly level desert floor. The beam is straight, but the desert floor curves 4.9 meters over an 8000-meter (8 km) tangent, as shown in the sketch below (certainly not to scale!).

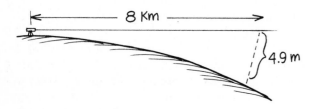

8 Km

4.9 m

After you have drawn this on the chalkboard, replace the laser with a super cannon positioned so it aims along the laser line. Ask your class how far along the laser line a cannonball would go if fired at 2 km/s, with no gravity and no air drag. Before you continue, be sure they see it will travel 2 km. Ask if this is fast enough to attain an earth orbit in the presence of earth gravity, with no air drag. [Its speed is insufficient for earth orbit. It will crash into the sand before it can reach the 2-km point.] Ask how far the cannonball will fall vertically beneath the laser line, providing the sand isn't in the way. [4.9 m] Draw the sketch below to show this case. You shovel sand out of the way if it is not to strike the earth's surface.

Next consider a cannonball fired at 4 km/s. Without gravity or air drag, it will travel 4 km down the laser path in 1 second. Ask if this is fast enough to attain an earth orbit. [No, it will hit the sand before 1 second is up. Emphasize that the cannonball fall, a vertical distance of 4.9 (or 5) m in 1 second, whatever its horizontal speed is.] So you dig some sand out of the way, and your sketch now looks like this:

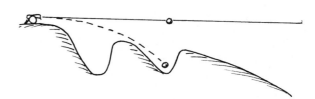

Continue by considering a greater muzzle velocity — great enough so that the cannonball travels 6 km in 1 second. This is 6 km/s. Ask if this is fast enough not to hit the sand (or equivalently, if it is fast enough to attain an earth orbit!) Then repeat the previous line of reasoning, again having to dig a trench. Now your sketch looks like this:

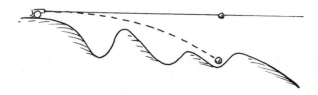

Continue by considering a still greater muzzle velocity — 8 km/s. Ask for a neighbor-check about this speed. Ask if you'll have to dig sand out of the way for this speed. Then after a pause and with a tone of importance, ask the class with what speed the cannonball must have to orbit the earth. Done properly, you have led your class into a "derivation" of orbital speed about the earth with no equations or algebra!

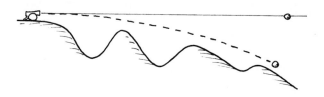

Acknowledge that gravitational force is less on satellites in higher orbits so they do not need to go so fast. This is acknowledged in the footnote on text page 194. (Since $v = \sqrt{GM/d}$, a satellite at 4 times the earth's radius needs to travel only half as fast, 4 km/s.)

Tangential Motion: Establish the idea that gravity does not change the tangential speed of a satellite — that there is no tangential component of gravitation. Do this by considering the effect of gravity on a bowling ball rolling along a level bowling alley. The pull is down, perpendicular to the alley and perpendicular to the direction of motion. Therefore, gravity does no work on the ball (text Figure 14-5). Consider a bowling alley that completely encircles the earth — elevated so it is above air drag. The ball would roll indefinitely — always "level."

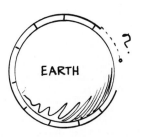

CHECK QUESTION: How fast would the bowling ball have to be moving for it to clear a broken span and continue moving along the alley on the other side? [8 km/s; In fact, you could remove the whole alley!]

CHECK QUESTION: Would a cannonball fired *upward* at 8 km/s go into an earth orbit? [No, it would simply act as a projectile and crash back into the earth at 8 km/s. To circle the earth it must have a *tangential* speed of 8 km/s.]

(This is a good place to break.)

Elliptical Orbits: Begin your treatment of elliptical orbits with the following demostration.

DEMONSTRATION [14-1]: With a loop of string and a pair of small suction cups stuck to the chalkboard, trace an ellipse as is done in text Figure 14-7.

Return to Newton's Mountain and consider greater cannonball speeds, starting with 9 km/s. Show how this speed causes the cannonball to overshoot the path it would take for a circular orbit. Ask if the 9 km/s value will increase, decrease, or remain the same on the first part of its outward trip. [Since it is going against gravity, the cannonball will slow to a speed less than its initial speed — quite a different situation than when in circular orbit.] Trace a full ellipse. As you retrace the elliptical path, show with a sweeping motion of your arm how the satellite slows as it recedes from the earth, moving slowest at its farthermost point. Then show how it speeds up as it falls toward the earth, whipping around the earth at its closest point. The cycle repeats. Point out the similarity of this to a stone which, when thrown upward at an angle, slows on the way up and speeds up on the way down. Planets orbiting about the sun do so in a similar way. Kepler didn't understand the slowness of planets when farthest from the sun, because he did not view them as bodies in free fall around the sun.

CHECK QUESTIONS: If a cannonball is fired horizontally from Newton's mountain at a tangential velocity less than 8 km/s, it soon strikes the ground below. Will its speed of impact be greater, the same, or less than its muzzle speed? [It will be greater, because of a component of its velocity is along the gravitational field of the earth. Similarly, any object tossed horizontally that moves downward will pick up speed for the same reason.] State that only if it is fired at 8 km/s will its speed remain at 8 km/s. If it is fired at 9 or 10 km/s, how does its speed change? [Its speed decreases, because a component of its velocity is against the gravitational field. Simply put, it is going against gravity.]

Summarize by stating that for an orbit close to earth, tangential satellite speeds must range between 8 km/s and 11.2 km/s.

Escape Speed: Toss something straight upward. Point out that if air drag does not play a role, then the launching speed and the speed of return are the same. Firing a projectile upward at

8 km/s will result in its having a return speed of 8 km/s if there is no air drag. The kinetic energy lost going up is equal to the kinetic energy gained in returning. But beyond 11.2 km/s, the story is different. This speed is sufficient for a no-return situation. This is escape speed.

CHECK QUESTION: If an object located at a distance beyond Pluto were dropped from a position of rest to earth, what would be its maximum speed of impact if its increase in speed is due only to earth gravity? [Interestingly, the max speed is 11.2 km/s, the same speed it would need to bounce from the earth and return to its original distance.]

Acknowledge that the term *escape speed* refers to "ballistic speed," the speed a body must have after the thrusting force ends. If the thrusting force were somehow continuous, then any speed could provide escape if maintained over sufficient time.

Energy Conservation: Sketch a large ellipse on the board to represent an elliptical orbit around the earth. Place the earth in the appropriate place. (It is invariably closer to the perigee than most people would place it — see the position of the focus in text Figure 14-8). Now place a satellite at the perigee and write a large "KE" beside it. That's where the satellite is traveling the fastest. But it's also closest to the earth, so write a small "PE" next to the "KE." Draw the satellite at other points. Ask for relative values of KE and PE at these points. Express these with the exaggerated-symbol technique. After discussion, erase the board.

CHECK QUESTION: Draw another ellipse on the board, with a planet appropriately placed. Sketch several satellite positions as shown and label them A, B, C, and D. Ask where the satellite has the maximum (1) speed, (2) velocity, (3) gravitational force to earth, (4) kinetic energy, (5) momentum, (6) gravitational potential energy, (7) total energy, and (8) acceleration. [(1) A; (2) A; (3) A; (4) A; (5) A; (6) C; (7) same everwhere; (8) A. You can expect the greatest number of wrong answers on the last question. This is where the guide for thinking, $a = F/m$, comes in handy.]

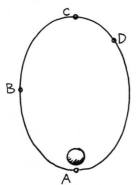

More Think-and-Explain Questions

1. If a projectile were launched from the surface of the earth at a vertical speed of 10 km/s, would it orbit the earth?

   Answer: No, it would simply rise and then fall back to earth, just as a vertically thrown baseball would do. It must have a horizontal component of 8 km/s to orbit — lesser speeds are required at greater distances from the earth.

2. If Pluto were somehow stopped dead in its orbit, it would fall into, rather than around, the sun. How fast would it be moving when it hits the sun?

   Answer: It would hit the sun with a speed very near the escape speed from the sun — 620 km/s.

3. What would be the consequences of the sun momentarily stopping in its apparent path across the sky?

   Answer: It would be the earth that stops, not the sun. With no tangential speed, the earth would then fall directly toward, rather than around, the sun.

4. The orbiting space shuttle moves at 8 km/s with respect to the earth. Suppose it projects a capsule rearward at 8 km/s with respect to the shuttle. Describe the path of the capsule with respect to the earth.

   Answer: Its tangential speed would be zero, and it would fall vertically to the earth's surface.

Computational Problems

1. The force of gravity between the earth and an earth satellite is given by $F = GmM/r^2$, where m is the mass of the satellite, M the mass of the earth, and r the radial distance between the satellite and the center of the earth. If it follows a circular orbit, the force is a centripetal force, given by $F = mv^2/r$. Equate the two expressions for force to show that the speed $v = \sqrt{GM/r}$.

   Answer: $GmM/r^2 = mv^2/r$; Cancelling m and r, we get $GM/r = v^2$; taking the square root of both sides gives $v = \sqrt{GM/r}$.

2. The speed of a satellite in circular orbit is given by the equation, $v = \sqrt{GM/r}$, where G is the gravitational constant, M is the mass of the earth, and r is the radial distance between the satellite and the center of the earth. Equate this to the other expression for speed, $v =$ distance traveled/time. Find the equation for the time it takes to completely orbit the earth — the period T. Use the complete orbit, $2\pi r$, for the distance traveled, and T for the period of revolution.

   Answer: $T = 2\pi\sqrt{r^3/GM}$. Begin with $v = \sqrt{GM/r}$ $= 2\pi r/T$; invert both sides and we have $\sqrt{r/GM}$ $= T/2\pi r$, and rearrange to find $T = 2\pi r\sqrt{r/GM}$. Since $r = \sqrt{r^2}$, this can be written $T = 2\pi\sqrt{r^3/GM}$.

15 Special Relativity – Space and Time

To use this planning guide work from left to right and top to bottom.

See the Chapter Notes for alternative ways to use all these resources.

Objectives

- Give examples of relative and non-relative motion.
- Define and give examples of Einstein's first and second postulate of Special Relativity.
- Give an example of time dilation.
- Reconcile the concept that observers in different frames of reference moving uniformly with respect to each other, each observe time dilation in the other frame by using the example of the twin trip, in which both twins agree that the traveling twin does not age as fast.
- Explain how a space traveler could live long enough to travel a distance which light takes 200 years to travel.

Possible Misconceptions to Correct

- Objects can go faster than the speed of light from some frames of reference.
- There is no upper limit of speed for material objects.
- Our rate of aging does not depend on the frame of reference.
- The farthest one can travel in a time of 1 year at the speed of light is 1 light year, regardless of the frame of reference.

No Demonstrations or Labs for this Chapter

Introduction

For those who will not be covering all the chapters of the text, I expect this chapter and the following one to be omitted. They can be omitted without consequence to the chapters that follow. The ideas discussed in these chapters are perhaps the most exciting in the book, but they are difficult to comprehend. Regardless of how clearly and logically this material is presented, students may find that they do not "understand" it in a manner that satisfies them. This is understandable for so brief an exposure to a part of reality untouched by conscious experience. The intention of these chapters is to develop enough insight into relativity to stimulate further student interest and inquiry .

Time is one of those concepts we are all familiar with and yet are hard pressed to define. A simple, yet less than satisfying, way to look at it is like our definition of space (that which we measure with a measuring stick) and time (that which we measure with a clock). Or we may quip, time is nature's way of seeing to it that everything doesn't happen all at once!

Should you get into a discussion of the relativity of time NOT as meant in the special theory, an interesting point to bring up is the longevity of different mammals — for example, the short life of a mouse compared to the longer life of an elephant. Based on the internal clocks of their own hearts or the rhythm of their own breathing, all mammals live about the same amount of time — about 200 million breaths and about 800 million heartbeats. So don't feel sorry for pets that, from our point of view, live for such short life spans. (The exception, interestingly or luckily enough, is humans, who have about three times as many heartbeats and breaths on the average than animals do.)

Note the important significance of *The Twin Trip*, Section 15.7 in the text, in that it completely bypasses mathematical equations for time dilation and the relativistic Doppler effect. The reciprocity of relativistic Doppler frequencies for approach and recession stems only from Einstein's first and second postulates. It is illustrated with a 4-step presentation that involves only simple arithmetic in the following suggested lecture.

This reciprocal nature of the relativistic Doppler effect does not hold for waves that require a medium, such as sound, where the "moving" frame is not equivalent to the "rest" frame (relative to air). If the ratio of the frequency received to the frequency sent for hearing in the rest frame is 2, the ratio for hearing in the moving frame is 3/2 (clearly not 2!). For sound, the speed as well as the frequency depend on the motion of the receiver. If the receiver moves toward a sound source, then the speed of sound encountered is greater. If moving away, it is less — very unlike the case for light.

With the simple flash-counting sequence, time dilation is shown without the use of any mathematical formulas. The results of the Twin-Trip flash sequence agree with Einstein's time dilation equation. So this treatment is completely independent of the time dilation equation and the relativistic Doppler equation! (Who says that good physics can't be presented non-mathematically?)

If your class is in a more mathematical mood, you may wish to show an alternative approach to the Twin Trip and consider straightforward time dilation, plus corrections for the changing positions of the emitting or receiving body between flashes. Instead of bypassing the time dilation equation, use it to show that at 0.6c, 6-minute flash intervals in the emitting frame compute to be 7.5-minute flash intervals in the receiving frame. The flashes would be seen at 7.5-minute intervals if the ship were moving crosswise (neither approaching nor receding),

such that each flash travels essentially the same distance to the receiver. In our case the ship doesn't travel crosswise, but it recedes from and then approaches the receiver. Consequently, corrections must be made in the time interval due to the extra distance the light travels when the spaceship is receding, and the lesser distance the light travels when the ship is approaching. This turns out to be 4.5-minutes.

$$\Delta t = \frac{\text{extra distance}}{c} = \frac{0.6c \times 7.5 \text{ min}}{c} = 4.5 \text{ min}$$

So when receding, the flashes are seen at 7.5 + 4.5, or 12-minute intervals. When approaching, the flashes are seen at 7.5–4.5, or 3-minute intervals. The results of this method are the same as those of the 4-step conceptual presentation in the following suggested lecture.

Suggested Lecture

After discussing Einstein and a broad overview of what special relativity is and is not, point out somewhere along the line that the theory of relativity is grounded in *experiment*, and in its development, it explained some very perplexing experimental facts (constancy of the speed of light, muon decay, solar energy, the nature of mass). It is not, as some people think, only the speculations of one man's way of thinking. Newspapers during the early part of the century used to report that there were only 12 people in the world who understood special relativity. This is inaccurate. For although in 1905 the only one person to understand special relativity was Einstein himself, after he published his paper and explained it, large numbers of people in the physics community also understood it.

Motion is Relative: Ask your class to pretend they are in a parking lot playing ball with someone who is driving toward them and away from them in an open vehicle (Figure 15-5, text page 207). A pitcher in the vehicle tosses a ball to them, always at the same pitching speed and without variation. Ask for the relative speed of catching a ball when the car approaches and again when it recedes. They know there will be a difference. Ask how they would react if the speed of the ball was always the same when caught, regardless of whether the thrower was moving toward them, at rest, or moving away from them. This would be most perplexing. State that a similar occurrence was presented to physicists at the turn of the century by the null result of the Michelson-Morley experiment.

Michelson-Morley Experiment: The experiment that first showed that the speed of light is invariant was the 1887 Michelson-Morley experiment. Avoid information overload by not treating the details of their experiment and the development of the interferometer. Instead, direct your students' mental energies to the broad ideas of special relativity. Explain what it means to say that the velocity of light is invariant — that it is the same for all observers.

First Postulate: The laws of physics are the same in all uniformly-moving reference frames. A bee inside a fast-moving jet plane executes the same flying maneuvers regardless of the speed of the plane. If you drop a coin to the floor of the moving plane, it will fall as if the plane were at rest. A flight attendant need make no adjustments in pouring tea because of the plane's high speed. Physical experiments behave the same in all uniformly-moving frames. This leads, most importantly, to the development of special relativity — to the speed of light, which is seen to be the same to all observers.

Second Postulate: Stand still and toss a piece of chalk in the air, catching it as you would when flipping a coin. Ask the class to suppose that, in so doing, all measurements show the chalk to have a constant average speed. Call this constant speed c for short. Then proceed to walk at a fairly brisk pace across the room and again toss the chalk into the air. State that from your frame of reference the measured speed is again the same. Ask if the speed looked different to them. They should respond that the chalk was moving faster this time. Ask

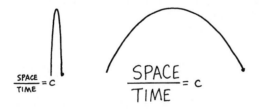

them to suppose instead that their measurement of speed was the same. They may be a bit perplexed, which again is similar to the perplexed state of physicists at the turn of the century. Write on the board, in uniformly-sized letters,

$$c = \frac{\text{SPACE}}{\text{TIME}}$$

This represents speed as seen by you in your frame of reference. State that from the frame of reference of the class, the space covered by the tossed chalk was seen to be greater. Write the word SPACE in correspondingly larger letters and

underline it. State that if they measure the same speed, that is, the same ratio of space to time, then such can be accounted for if the measured time is also greater. Then write the enlarged word TIME beneath the underline, equating it to c. Just as π, the ratio of circumference to diameter, is the same for all sizes of circles, all observers similarly measure the same ratio of space to time for light waves in free space.

Relate the analogy of your chalk-tossing sequence to the light clock discussed on page 211, in Figures 15-11 through 15-14, and in the boxed material on page 213 in the text.

Relativistic Doppler Effect: On the board, sketch a simple version of Figure 4 (shown on the next page). Explain that for a ship *at rest*, relative to the two observers on the distant planets, light flashes emitted at 6-minute intervals would be seen by both to be at 6-minute intervals also. But with motion, the situation is different. Give two examples of the Doppler effect; the changing pitch of a car horn when it approaches and when it recedes, and the pitter-patter of a slanting rain when you run into the rain versus when you run away from it. Ask your class to suppose that the ship moves so fast toward the right observer that the flashes reach the observer at twice the frequency — with the flashes closer together so they appear at 3-minute intervals. The time between flashes is crimped in half. Ask how the time between flashes would be seen by the observer on the left, who sees the source receding. Is it reasonable to say the opposite occurs? That is, instead of being crimped in half, the flashes are spread apart twice as much, so that the time between flashes is stretched by two? Or, if 6-minute flash intervals are crimped to 3 minutes for approach, they'll be stretched to 12-minute intervals for recession? If this is acceptable to your class, you can then go on to discuss the Twin Trip, as is done in the text. Depending on the level of your class, you may wish to derive this halving and doubling of time intervals with the method below.

Option: Derivation of Relativistic Doppler Effect: Before treating the light-flash sequence of The Twin Trip on text pages 214 -220, you may want first to establish the reciprocal relationship between approaching and receding frequencies — that is, the relativistic Doppler effect. The conventional way to do this in mathematically-oriented physics classes is to derive algebraically the expression

$$f = f_0 \sqrt{\frac{1 + v/c}{1 - v/c}}$$

This derivation can be found in many physics texts. However, you can derive the same result without using a single mathematical expression and show that the reciprocity of frequencies is a natural consequence of the invariance of the speed of light. Try this with the following 4-step conceptual presentation.

Step 1: Consider a person standing on earth and directing brief flashes of light at 3-minute intervals to a distant planet at rest relative to the earth. Some time will elapse before the first of these flashes reaches the planet, but since there is no relative motion between the sender and receiver, successive flashes will be observed on the distant planet at 3-minute intervals. While you are making these remarks, draw a sketch of Figure 1 on the board.

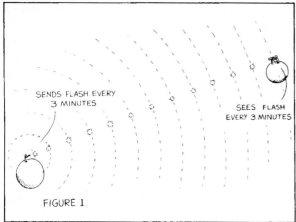

FIGURE 1.

Step 2: How frequently would these flashes encounter an observer in a fast-moving spaceship traveling between the earth and the planet? Although the speed of the flashes measured by the spaceship would be *c*, the *frequency* of flashes would be greater than or less than the emitting frequency, depending on whether the ship was receding or approaching the light source. After supporting this idea with some examples of the Doppler effect (car horns, running into versus away from a slanting rain), make the supposition that the spaceship recedes from the light source at a speed that is great enough for the frequency of the light flashes to decrease by half, so they're seen from the ship only half as often, at 6-minute intervals. By now your chalkboard sketch looks like Figure 2.

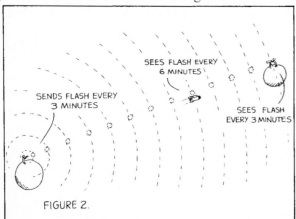

FIGURE 2.

Step 3: Now suppose that each time a flash reaches the ship, a triggering device activates a beacon on the ship that sends its own flash of light toward the distant planet. According to a clock in the spaceship, this flash is emitted every 6 minutes. Since the flashes from earth and the flashes emitted by the spaceship travel at the same speed, *c*, both sets of flashes travel together and an observer on the distant planet sees not only the earth flashes at 3-minute intervals, but the spaceship flashes at 3-minute intervals as well (Figure 3). At this point, you have established that 6-minute intervals on the approaching spaceship are seen as 3-minute intervals on the stationary planet.

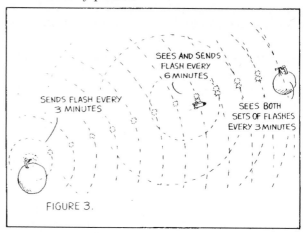

FIGURE 3.

Step 4: To establish that the 6-minute flashes emitted by the spaceship are seen at 12-minute intervals from the earth, go back to your earlier supposition that 3-minute intervals on earth are seen as 6-minute intervals from the frame of reference of the receding ship. Pose this question to your class. If instead of emitting a flash every 3 minutes, the person on earth emits a flash every 6 minutes. How often would these flashes be seen from the receding ship? [12 minutes] Next ask if the situation would be any different if the ship and earth were interchanged, that is, if the ship were at rest and emitted flashes every 6 minutes to a receding earth? [No] After a suitable response, erase from your chalkboard drawing all the flashes emitted from the earth. Replace the earth-twin's light source with a telescope, while asking how often the 6-minute flashes emitted by the moving spaceship are seen from earth. Student response should show that you have established the reciprocity of frequencies for the relativistic Doppler effect without using a single equation. This is summarized in Figure 4.

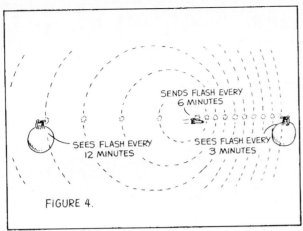

FIGURE 4.

Note that you have employed Einstein's postulates in the last 2 steps, the second postulate in Step 3 (constancy of the speed of light) and the first postulate in Step 4 (equivalence of the earth and ship frames of reference).

Whether you have established this reciprocity from the Doppler equation or from the preceding four steps, you are now ready to demonstrate time dilation while also presenting a resolution to the so-called twin paradox. Do this by reviewing the treatment The Twin Trip in text Section 15.7.

Applications of Time Dilation: Discuss the prospects of "century hopping," a scenario in which future space travelers may take relatively short trips of a few years or so, and return decades, centuries, or even thousands of years later. Of course this depends on the solution to two major problems (1) sufficient rocket engines and fuel supplies for prolonged voyages, and (2) a means of shielding from the radiation that would be produced by impact with interstellar matter.

Present the following interesting, but fictitious, example of time dilation. Suppose that one could be whirled in a giant centrifuge up to relativistic speeds without physical injury. (Of course in reality one would be crushed to death in such a case, but pretend that somehow one is physically unaffected by the crushing centripetal forces —hence the fictitiousness of this example.) Then cite how one taking a "ride" in such a centrifuge might be strapped in a seat and told to press a button on the seat when he or she wishes the ride terminated. Now suppose that, after being whirled about at rim speeds near the speed of light, the occupant decides 10 minutes is long enough. So he or she presses the

button, signaling those outside to bring the machine to a halt. After the machine is halted, those outside open the door peer in and ask, "Good gosh, what have you been doing in there for the past 3 weeks!" In the laboratory frame of reference, 3 weeks would have elapsed during a ten-minute interval in the rotating centrifuge. The point: One doesn't necessarily have to travel through wide expanses of space for time dilation to be significant. Motion in space, rather than space itself, is the key factor.

More Think-and-Explain Questions

1. Could a person who has a life expectancy of 70 years possibly make a round-trip journey to a part of the universe thousands of light years distant?

   Answer: If a person travels at relativistic speeds, distances as far as those that light takes thousands of years to travel (from our frame of reference) could be traversed well within an average lifetime. This is because distance is relative to the frame of reference from which it is measured. Long distances from a rest frame may be quite short from a moving frame.

2. A twin who makes a long trip at relativistic speeds returns younger than his stay-at-home twin sister. Could he return before his twin sister was born? Defend your answer.

   Answer: A twin who makes a long trip at relativistic speeds returns younger than his stay-at-home twin sister only in the sense that he has not aged as much as his sister during the duration of his trip. If they could watch each other during the trip, at no time would either see a reversal of aging. They would see only a slowing down or speeding up of aging processes. A reversal would appear only for speeds greater than the speed of light. So aging processes will slow down, not reverse. This means the twin could not return to a time before his birth or that of his twin sister.

3. Is it possible for a person to be biologically older than one's parents? Explain.

   Answer: Yes, it is possible for one to be biologically older than one's parents. Suppose, for example, that a woman of any age gives birth to a baby and then departs in a high-speed rocket ship. She could return from a relativistic trip in a few years to find her baby is 50 years or so older than when she left.

Special Relativity – Length, Energy, and Momentum

To use this planning guide work from left to right and top to bottom.

Chapter 16 Planning Guide

• *The bulleted items are key: Be sure to do them!*

Topic	Exploration	Concept Development	Application
Length Contraction		• Text 16.1/Lecture	
Mass-Energy Relationship		• Text 16.2/Lecture Con Dev Pract Pg 16-1	Nx-Time Q 16-1
Relativistic Momentum		Text 16.3/Lecture	
Relativistic Kinetic Energy		Text 16.4/Lecture	
Correspondence Principle		• Text 16.5/Lecture	

Video: *Special Relativity 2*
Evaluation: Chapter 16 Test

See the Chapter Notes for alternative ways to use all these resources.

Objectives

After studying Chapter 16, students will be able to:
• Describe the conditions under which lengths contract.
• Describe the mass-energy relationship.
• Correctly interpret the equation $E_0 = mc^2$.
• Explain how mass conservation and energy conservation hold true despite the equivalence of mass and energy.
• Explain why the equivalence of mass and energy is not noticed for everyday events, such as the throwing of a baseball.
• Explain how the correspondence principle is a good test of the validity of any new theory.

Possible Misconceptions to Correct

• Objects can go faster than the speed of light in some frames of reference.
• $E_0 = mc^2$ means that energy is mass traveling at the speed of light squared.
• The momentum of an object is always simply its mass x velocity.

No Demonstrations for this Chapter

Introduction

This chapter is a continuation of the previous chapter. It (or both relativity chapters) may be omitted from your course without impacting the chapters that follow. Mass-energy is treated again in Chapter 40. If you omit this chapter, bring up the mass-energy concept at that time.

As stated in the text, push an object that is free to move and it accelerates in accordance with Newton's second law, $a = F/m$. Interestingly enough, Newton originally wrote the second law not in terms of acceleration, but in terms of momentum. The momentum form, $F = \Delta p/\Delta t$, is equivalent to the familiar $F = ma$. (Not until this chapter do we assign the symbol p to momentum.

It wasn't introduced in Chapter 7 in order to avoid presenting another abstract symbol to the student's early vocabulary. It wouldn't have simplified anything then, but it is useful. Hence we say $p = mv$, or when high speeds are involved, $p = mv/\sqrt{1 - (v^2/c^2)}$.

In accordance with the momentum version of Newton's second law, if we push an object that is free to move, we increase its momentum. The acceleration, or change-of-momentum, version of the second law gives the same result. But for very high speeds, it turns out that the momentum version is accurate while $F = ma$ is incorrect. $F = \Delta p/\Delta t$ holds for all speeds, from speeds you are familiar with everyday to speeds near the speed of light — as long as the relativistic expression for p is used.

The first edition of *Conceptual Physics* spoke of *relativistic mass*, given by the equation $m = m_0/\sqrt{1 - (v^2/c^2)}$. The concept considered was that the mass of an object increases as energy is supplied to it. This idea, expressed in many textbooks, is now losing ground in favor of the somewhat more complex idea of relativistic momentum. One problem with the idea of relativistic mass is that mass is a scalar; it has no direction. When a particle is accelerated to high speeds, its apparent increase in mass is directional. The increase in mass occurs in the direction of motion in a manner similar to the way that length contraction occurs only in the direction of motion. But moving mass is, after all, momentum. So it is more appropriate to speak of increases in momentum rather than increases in mass. Both treatments, however, describe rapidly-moving objects in accordance with observations.

Suggested Lecture

Length Contraction: Hold up a meter stick horizontally. State that if your students made an accurate measurement of the meter stick's length, their measurement would agree with your own. Everyone would measure the meter stick to be one meter long. People at the back of the room would have to compensate for its shorter appearance due to distance, but nevertheless, they would agree on its one-meter length. Now walk across the room holding the meter stick like a spear. State that your measurements and those of your students would now differ. If you were to travel at 87% the speed of light relative to the class, they would measure the stick to be half as long, or 0.5 m. At 99.5% the speed of light, they would see it as only 10 cm long. At greater speeds, the meter stick would be even shorter. At the speed of light, it would contract to zero length. Write the length-contraction formula on the board:

$$L = L_0\sqrt{1 - (v^2/c^2)}$$

State that contraction takes place only in the direction of motion. When the stick is moving in spear fashion, it appears shorter but it doesn't appear thinner.

Contrast the class's view of the stick with your view of it. Since you move with the stick you see no contraction, whatever the stick's speed. From your frame of reference (which is the same as the stick's), the v in the above equation is zero and $L = L_0$. Consequently, contraction depends on the frame of reference.

Consider traveling past a one-kilometer long building at very high speed. From your frame of reference in the traveling vehicle, the length of the building is shorter. At 87% the speed of light, for example, the building would be seen as half as long, 0.5 km. The same is true in space travel between stars. Their distance as seen from our frame of reference at rest is quite different than as seen from the frame of reference of a moving spaceship. If twenty light-years separate a pair of stars from our frame of reference, a spaceship traveling at 0.087c between them would see the stars as only ten light-years apart.

CHECK QUESTION: Consider a pair of stars, one on each "edge" of the universe. That's an enormous distance of separation from our frame of reference. Now consider a photon traveling from one star across the entire universe to the other. From the frame of reference of the photon, what is the distance of separation between stars? (How big is the universe?) [Zero! From a frame of reference traveling at c, the length contraction reaches zero.]

The implication of the above question is that at high speeds future space travelers may not face the restrictions of traveling distances that seem formidable without relativity! There is much food for thought provided here!

The Mass-Energy Relationship: On the board write $E_0 = mc^2$, the most celebrated equation of the twentieth century. It relates energy and mass. Every material object is composed of energy, "energy of being." This "energy of being" is appropriately called rest energy, which is designated by the symbol E_0. (We distinguish here between rest energy E_0 and the total energy E, which may include potential and kinetic energies as well as rest energy.) So the mass of something is actually the energy within it. This energy can be converted to other forms, light for example.

The amount of matter that is converted to radiant energy by the sun every second is 4.5 million tons. That tonnage is carried by the radiant energy through space. So when we speak of matter being "converted" to energy, we are merely converting it from one form to another—from a form with one set of units to perhaps another. Because of the mass and energy equivalence, the total amount of mass plus energy does not change in any reaction that takes the whole system into account.

Discuss the interesting idea that mass, every bit as much as energy, is delivered by the power utilities through the copper wires which run from the power plants to consumers.

From a distant perspective, it seems likely that the twentieth century will be seen as a major turning point in physics that occurred due to the discovery of the $E_0 = mc^2$ relationship. It may be interesting to speculate on what the equation of the twenty-first century might be.

Relativistic Momentum:

State that if you push an object that is free to move, it accelerates in accordance with Newton's second law, $a = F/m$. Newton originally wrote the second law not in terms of acceleration, but in terms of momentum, $F = \Delta p/\Delta t$. This form is equivalent to the familiar $F = ma$. Here we introduce the symbol p for momentum, $p = mv$. In accordance with the momentum version of Newton's second law, if we push an object that is free to move, we increase its momentum. The acceleration, or the change-of-momentum, version of the second law gives the same result. However for very high speeds, it turns out that the momentum version is more accurate. $F = \Delta p/\Delta t$ holds for all speeds, from the everyday familiar speeds to those near the speed of light—as long as the relativistic expression of p is used.

Write the expression for relativistic momentum on the board: $p = mv/\sqrt{1-(v^2/c^2)}$. Point out that it differs from the classical expression for momentum by the denominator $\sqrt{1-(v^2/c^2)}$. A common (though not completely accurate) interpretation is that of a relativistic mass $m = m_0/\sqrt{1-(v^2/c^2)}$ multiplied by a velocity v. Because the "increase in mass" with speed is directional (as is length contraction) and momentum rather than mass is a vector, the concept of an increase in momentum rather than an increase in mass is preferred in advanced physics courses. The treatments of relativistic mass and relativistic momentum, however, lead to the same description of rapidly-moving objects in accordance with observations.

A good example of the increase of either mass or momentum for different relative speeds is the accelerated electrons and protons in high-energy particle accelerators. In these devices, speeds greater than $0.99\ c$ are attained within the first meter, and most of the energy given to the charged particles during the remaining journey goes into increasing mass or momentum. The particles strike their targets with masses or momenta that is thousands of times greater than Newtonian physics would predict. Interestingly enough, if you traveled along with the charged particles, you would note no such increase in the particles themselves (the v in the relativistic mass equation would be zero), but you would measure a mass or momentum increase in the atoms of the "approaching" target. The crash is the same whether the elephant hits the mouse or the mouse hits the elephant.

Cite how such an increase must be compensated in the design of circular accelerators, such as cyclotrons, bevatrons, and the like, and how such compensation is not required for a linear accelerator, except for the bending magnets at its end.

Point out to your class that the form of the relativistic momentum equation is very much like that for time.

Show how for small speeds the relativistic momentum equation reduces to the familiar mv (just as for small speeds $t = t_0$ in time dilation). Then show what happens when v approaches c. The denominator of the equation approaches zero. This means that the momentum approaches infinity! An object pushed to the speed of light would have infinite momentum and would require an infinite impulse (force × time). This is clearly impossible. Nothing material can be pushed to the speed of light. The speed of light c is the upper speed limit in the universe.

Cars, planes, and even the fastest rockets don't approach speeds to merit relativistic considerations, but subatomic particles do. They are routinely pushed to speeds beyond 99% the speed of light, whereupon their momenta increase thousands of times more than the classical expression mv predicts. This is evidenced when a beam of electrons directed into a magnetic field deflects less than classical physics predicts. The greater its speed is, the greater is its "moving inertia" (its momentum), and the more it resists deflection (text Figure 16-6). Physicists studying high-energy physics must take relativistic momentum into account when working with high-speed subatomic particles in atomic accelerators. In that arena, relativity is an everyday fact of life.

Relativistic Kinetic Energy: The principal value of this section is to answer the question, "From what did Einstein's equation $E_0 = mc^2$ originate?" Einstein was the first to derive the relativistic expression for kinetic energy, KE $= mc^2/\sqrt{1 - v^2/c^2} - mc^2$ and the first to note the term mc^2, which is independent of speed. This term is the basis of the celebrated equation, $E_0 = mc^2$.

Interestingly enough, for ordinary low speeds the relativistic equation for kinetic energy reduces to the familiar KE $= 1/2\ mv^2$ (via the binomial theorem). As the footnote on page 231 suggests, in many situations where the momentum or energy, rather than the speed, of high-speed particles is known, the expression that relates total energy E to the relativistic momentum p is given by the formula $E^2 = p^2c^2 + (mc^2)^2$. This expression is derived by squaring the relation $E_0 = mc^2$ to obtain $E^2 = m^2c^4 = m^2c^2(c^2 + v^2 - v^2)$, and combining the relativistic equation for momentum.

Like the argument for a speed limit via the infinite impulse required to produce infinite momentum, we find that doing more and more work to move an object increases its kinetic energy disproportionately compared to its increase in speed. The accelerated matter requires more and more kinetic energy for each small increase in speed. An infinite amount of energy would be required to accelerate a material object to the speed of light. Since an infinite amount of energy is not available, we again conclude that material particles cannot reach the speed of light.

The Correspondence Principle: This is one of the neatest principles of physics. It is also a guide to clear and rational thinking, not only about the ideas of physics but for all good theory, even areas as far removed from science as government and ethics. Simply put, if a new idea is valid then it ought to be in harmony with the region of established theory it overlaps.

Show your students that when small speeds are involved, the relativity formulas reduce to the everyday observation that time, length, and the momenta of things do not appear any different when they are moving. That's because the differences are too tiny to detect.

More Think-and-Explain Questions

1. As a meter stick with a normal mass of 1 kg moves past you, your measurements show it to have a mass of 2 kg. If your measurements show it to have a length of one meter, what is the orientation of the stick?

   Answer: The stick must be oriented in a direction perpendicular to its motion, unlike that of a properly-thrown spear. This is because the stick is traveling at relativistic speed (actually 0.87c), as evidenced by its increase in mass. The fact that its length is unaltered means that its long direction is not in the direction of motion. The thickness or height of the stick, not the length of the stick, will appear shrunken to half size.

2. In the preceding question, if the stick is moving in a direction along its length (as a properly-thrown spear would), how long will it appear to you?

   Answer: The stick will appear to be one-half meter long when it moves with its length along the direction of motion. Why one half its length? It is moving fast enough (0.87c) for its mass to be doubled.

3. How does the measured density of a body compare when the body is at rest and when it is moving?

   Answer: The density of a moving body is measured to increase because of a measured increase in mass and decrease in volume.

4. What does the formula $E_0 = mc^2$ mean?

   Answer: $E_0 = mc^2$ means that energy and mass are equivalent to each other. The c^2 is the proportionality constant that links the units of energy and mass. In a practical sense, energy and mass are one and the same. When something gains energy, it gains mass. When something loses energy, it loses mass. Mass is simply congealed energy.

5. *Muons* are elementary particles that are formed high in the atmosphere by the interactions of cosmic rays with gases in the upper atmosphere. Muons are radioactive and have an average lifetime of about two millionths of a second. Even though they travel at almost the speed of light, they are so high that very few should be detected at sea level— at least according to Newtonian physics. However, laboratory measurements show that muons in numbers of great proportions *do* reach the earth's surface. Can this be explained?

   Answer: This can be explained by time dilation or by length contraction. Time dilation: At their high speeds, muons have about ten times as much time, or twenty millionths of a second, to live. From the frame of reference of the stationary earth, the muon's "clock" is running ten times slower than the earth's clock, thus allowing sufficient time to make the trip. Length contraction: From the muon's frame of reference, the distance to earth is contracted ten times, so the muon has sufficient time to get there.

The Atomic Nature of Matter

To use this planning guide work from left to right and top to bottom.

Chapter 17 Planning Guide

• *The bulleted items are key: Be sure to do them!*

Topic	Exploration	Concept Development	Application
Atoms		• Text 17.1–17.4/Lecture	
Molecules	• Act 41 (1 period)	• Text 17.5/Lecture Con Dev Pract Pg 17-1	Exp 42 (>1 period)
Compounds		Text 17.6/Lecture	
Atomic Nucleus		• Text 17.7, 17.8/Lecture	Nx-Time Q 17-1, 17-2
States of Matter		• Text 17.9/Lecture	
Video: *Atoms* **Evaluation:** Chapter 17 Test			

See the Chapter Notes for alternative ways to use all these resources.

Objectives

After studying Chapter 17, students will be able to:

• Describe the relationship between atoms and elements.

• Compare the ages of atoms to the ages of the materials they compose.

• Give examples that illustrate the small size of atoms.

• Cite evidence for the existence of atoms.

• Distinguish between an atom and a molecule.

• Distinguish between an element and a compound.

• Identify the three basic building blocks that make up an atom and tell where in the atom each is found.

• Explain the significance of the horizontal rows and the vertical columns in the periodic table.

• Describe how the plasma state of matter differs from the solid, liquid, and gaseous states.

Possible Misconceptions to Correct

• Material things are made of thousands of different kinds of atoms.

• The atoms that make up a newborn baby were made in the mother's womb.

• The age of atoms in a baby is less than the age of atoms in an old person.

• Molecules are actually big atoms.

No Demonstrations for this Chapter

Introduction

The treatment of atoms in this chapter is basic and provides a good background for the chapters on heat. It also provides background for Chapters 28, 32, and Unit VI. The subject of atoms is treated further in Chapters 38 and 39. This chapter is the most important chapter in Unit II, and it should not be skipped.

The following interesting information is not in the text. The ten most abundant elements in our environment are, in order; oxygen, silicon, aluminum, iron, calcium, sodium, potassium, magnesium, titanium, and hydrogen. There are 4×10^{-6} grams of gold in a metric ton of seawater.

You may want to use the excellent 10-minute film *Powers of Ten* by Charles and Ray Eames with narration by Philip Morrison (Pyramid Films, 1978). This film makes an excellent comparison of the atom to the solar system, to galaxies, and to the universe and back to the atom, in terms of their sizes as positive and negative powers of ten.

Suggested Lecture

Begin by posing the scenario of breaking a boulder into rocks, rocks into gravel, gravel into sand, sand into powder, and so forth until you get to the fundamental building block of matter—the atom. Use examples to convey the smallness of the atom. For instance, an atom is as many orders of magnitude smaller than a person as the average star is larger than a person—so we stand somewhere between the atoms and the stars. The size of an atom is to the size of an apple as the size of the apple is to the size of the earth. So if you want to imagine an apple full of atoms, think of the earth solid-packed with apples.

Elements: State that although the number of atoms that exists is enormous, the number of different kinds of atoms is about 100. These are the elements from the simplest and lightest, hydrogen, to the most complex and heaviest normally found in the earth's crust, uranium. Elements heavier than uranium are not normally found in the earth's crust. They are the "transuranic elements."

Smallness of Atoms: You can use the following example to lead into the idea that there are more molecules in the air in your lungs than there are lungsful of air in the world. State that if you put a drop of ink in a bathtub full of water, very soon you will be able to sample any part of the water and find ink in it. The atoms of ink spread out. We can get an idea of the smallness of atoms from the following fact. There are more atoms in a thimbleful of ink than there are thimblefuls of water in all the lakes and rivers of the world. Consequently, if you throw a thimbleful of ink into one of the Great Lakes, eventually it will mix, and if you dip anywhere in the lake with a thimble, you'll have many atoms of ink in your sample.

Atoms are so tiny that you inhale billions of trillions with each breath, nearly a trillion times more atoms than the total population of people since humans emerged. [There are about 10^{23} atoms in a breath (one liter), and the present human population of the world is about 10^9, probably no more than 10^2 times the human population since time zero.] Assuming that most of the atoms previously breathed by people are still part of the atmosphere, you inhale billions of atoms which were exhaled by every person who ever sighed, sneezed, laughed, and simply breathed! So in this sense we are all one!

About six years are required for molecules released in the atmosphere to become uniformly mixed. Sneeze and then in six years travel anywhere in the world and inhale. More than likely, one of the molecules from the sneeze will be in that breath of air!

CHECK QUESTION: When you die, what ultimately becomes of the atoms of which you are made? [The atoms that make up your body and become part of the atmosphere will be breathed by and incorporated into the bodies of everyone else on earth—for all future generations!]

Cite how the process of cremation quickens the process of getting your atoms into the life cycle.

Historical Notes: Relate how from the earliest days of science people wondered how far the idea of breaking boulders into rocks, gravel, sand, and powder would go. Does the process of subdividing ever end? Hundreds of years ago people had no way of finding out, so they carried on with philosophical speculation. Not until the late 1700s did people begin to get indirect evidence of some basic order in the combinations of things. The first real "proof" for atoms was given by Einstein in 1905, the same year he published his paper on relativity. He calculated what kind of motion ought to compose Brownian motion. His calculations were based on ideas like energy and momentum conservation and on the idea of heat as atomic motion. Many leading scientists during that era didn't believe atoms existed until Einstein's work.

Evidence for Atoms: Ask what an atom would "look like" if viewed through a vertical bank of forty high-powered optical microscopes stacked one atop the other. Atoms don't *look* like anything—they don't have an appearance. You can qualify this by stating that they have no appearance in the range of frequencies we call light. Discuss text Figures 17-3 and 17-4. The historical significance of the Figure 17-3 photograph of individual atoms is that it has opened new doors in the fields of medicine and biology. No longer will guesswork be used to determine the positions of atoms in complex molecules.

You might allude to the later study of Chapter 38 and state that the electron beam in the electron microscope has the properties of high-frequency light. (The electron microscope that had its heyday more than three decades ago is a fundamentally different device than today's scanning electron microscopes and scanning tunneling microscopes.) Acknowledge the wave nature of matter—the fuzziness in the distinction between particles and waves at the atomic level. At the atomic level, particles also seem to be congealed standing waves of energy—"wavicles."

Molecules: Distinguish between atoms and molecules. There is a limited number of different atoms, but there are innumerable different molecules—and more are being discovered and constructed.

> CHECK QUESTION: What is the number of elements in a water molecule? What is the number of atoms in a water molecule? [Two elements (hydrogen and oxygen) and three atoms (two of hydrogen and one of oxygen)]

Interestingly enough, whereas an individual atom cannot be seen by the naked eye, some molecules can be seen. One such molecule, called a *macromolecule*, is a diamond. A diamond is actually one big carbon molecule!

The Atomic Nucleus: Distinguish among protons, neutrons, and electrons. Discuss Rutherford's discovery of the nucleus, the Bohr model of the atom, and the electrical role of the nucleus and surrounding electrons. Stress the emptiness of the atom and lead into the idea that solid matter is mostly empty space. State how our bodies are 99.999% empty space, and how a particle, if tiny and not affected by electrical forces, could be shot straight through us without even making a hole! Neutrons do just that. In a beam of neutrons aimed at the body, a few may make bull's-eye collisions with some of atomic nuclei—this can do damage so we wouldn't want to do this really. However, all but a minute fraction of neutrons in the beam would pass unhindered through the body.

Discuss the role of electrical forces that prevent us from oozing into our chairs. Ask the class to imagine that the lecture table is a large magnet and that you wear magnetic shoes that are repelled by the table you "stand" on. Then state that on the submicroscopic scale, this is indeed what happens when you walk on any solid surface. Only the repelling force isn't magnetic, it's electric! Discuss the submicroscopic notion of things touching. Acknowledge that under very special circumstances the nucleus of one atom can actually touch the nucleus of another atom. This is what happens in a thermonuclear reaction.

> CHECK QUESTION: True or False: There exists a large air gap between the nucleus of an atom and the orbiting electrons. [False, there is a void but not an air gap. Air is far

from being a void. It is a substance that consists principally of nitrogen and oxygen molecules.]

Schematically show the hydrogen atom, add a proton and neutrons to build a helium atom, and then add a lithium atom, and so on. Discuss atomic number and the role that the number of protons play in the nucleus in dictating the surrounding electron configuration. Call attention to and briefly discuss the periodic table on text page 247. Point out that the atomic configurations depicted in text Figure 17-10 are simply models. Models are not complete or accurate. For example, if the nuclei were drawn to scale they would be scarcely-visible specks. Also, the electrons don't actually orbit like planets as the drawings suggest—such terms don't seem to have much meaning at the atomic level. Electrons don't exactly orbit, because we lose the distinction between particles and waves. Can we say they "swarm" or "smear?" Your students who continue to study physics will eventually be up against the concept that something can be both a particle and a wave.

You might state that the configuration of electrons and their interactions with each other are the basis of chemistry.

States of Matter: Briefly discuss the states of matter and how different molecular speeds account for the solid, liquid, gaseous, and plasma states.

Quarks: Just as the nucleus is composed of protons and neutrons, the protons and neutrons themselves are composed of quarks. Are quarks composed of still smaller particles? There doesn't seem to be reason to think so, but we don't know. This uncertainty is often cited as a weakness by people who do not understand what science is about. Science is not a bag of answers to all the questions of the world, but rather it is a process for finding answers to many questions about the world. We continue to refine our models and add new layers to our understanding—sometimes building onto layers and other times replacing layers. It is unfortunate that some people see this as a weakness. This is remindful of Bertrand Russell, who publicly changed his mind about certain ideas during the course of his lifetime. These changes were part of his growth, but they were looked upon by some as a sign of weakness, as we discussed back in Chapter 1. The same is true in physics. Our knowledge grows, and that's nice!

More Think-and-Explain Questions

1. How likely is it that at least one of the atoms exhaled by your very first breath at birth will be in your next breath?

   Answer: It's highly likely, because the number of atoms exhaled is about the same as the number of liters of atmosphere in the world. Assuming you're more than six years old, they are uniformly mixed by now. However, that first breath was less than a liter, so take two or three breaths to be sure!

2. Which would produce a more highly valued element, adding or removing a proton from the element gold?

   Answer: Removing a proton: Gold turns into more-valuable platinum. Adding a proton turns gold into less-valuable mercury.

Computational Problems

1. How many grams of oxygen (O) are there in 18 grams of water?

   Answer: There are 16 grams of oxygen in 18 grams of water. We can see from the formula for water, H_2O, that there are twice as many H atoms (each with the atomic mass 1) as O atoms (each with the atomic mass 16). So the molecular mass of H_2O is 18, with 16 parts oxygen by mass.

2. How many grams of H are there in 16 grams of methane (CH_4) gas?

   Answer: A carbon atom is 12 times as massive as a hydrogen atom is, or three times as massive as 4 hydrogen atoms are. A bit of reasoning will show that for every 4 grams of hydrogen there will be 3×4, or 12 grams of carbon, which totals 16 grams of CH_4. So, there are 4 grams of hydrogen in 16 grams of methane.

3. Gas A is composed of diatomic molecules (2 atoms to a molecule) of a pure element. Gas B is composed of monatomic molecules (1 atom to a "molecule") of another pure element. Gas A has 3 times the mass of an equal volume of Gas B at the same temperature and pressure. How do the masses of elements A and B compare?

   Answer: The mass of element A is 3/2 the mass of element B. Why? Gas A has three times the mass of Gas B. If the equal number of molecules in A and B had equal numbers of atoms, then the atoms in Gas A would simply be three times as massive. However, there are twice as many atoms in A, so the mass of each atom must be half of three times as much, or 3/2.

18 Solids

To use this planning guide work from left to right and top to bottom.

Chapter 18 Planning Guide
• *The bulleted items are key: Be sure to do them!*

Topic	Exploration	Concept Development	Application
Crystal Structure		Text 18.1/Lecture	
Density		• Text 18.2/Lecture	
Elasticity		• Text 18.3/Lecture Demos 18-1, 18-2	• Exp 43 (1period)
Compression and Stretching		Text 18.4/Lecture Demo 18-3	
Scaling	Act 44 (>1 period)	• Text 18.5/Lecture Demos 18-4, 18-5 Con Dev Pract Pg 18-1, 18-2	Nx-Time Q 18-1

Video: *Scaling*
Evaluation: Chapter 18 Test

See the Chapter Notes for alternative ways to use all these resources.

Objectives

After studying Chapter 18, students will be able to:

- Cite evidence to show that many solids are crystals.
- Define density and explain why it is the same for different volumes or masses of the same material.
- Distinguish between an elastic material and an inelastic material.
- Predict the stretch for an applied force, given the stretch produced by a different force.
- Explain why the center of a horizontal steel girder need not be as wide as the top and bottom.
- Explain why making something larger by the same factor in all dimensions changes its strength in relation to its weight.

Possible Misconceptions to Correct

- Density is the same as weight, but it is expressed in different units.

- When a structure is scaled up or down in size, its properties go up or down in direct proportion.
- Doubling the volume of an object means doubling the surface area, or halving the volume means halving the surface area. Surface area and volume scale up or down in direct proportion.
- An animal (like the gorilla enlarged as King-Kong) scaled up in exact proportion would be proportionally stronger.

Demonstration Equipment

- [18-1] Two springs (a thick, stiff one to show compression and a thinner one to show elongation), stand, and weights to demonstrate Hooke's law
- [18-2] Balls of different materials to drop on a hard surface (such as an anvil) to show elasticity
- [18-3] Paper or plastic drinking straws and a potato
- [18-4] Spherical flask and cylindrical flask of equal volumes (500-ml or 1000-ml)
- [18-5] Eight cubes (plastic foam, wood, or plastic)

86

Introduction

This chapter may be skipped with no particular impact on other chapters. If you omit this chapter and assign Chapter 19, however, the concept of density should be introduced.

The crystal nature of solids is treated very briefly.

Hooke's law is treated in regard to elasticity and the stretching and compression of solids. In many courses Hooke's law is treated in mechanics. If you wish to do so, it can go with the material in Chapters 4 or 6.

Density is introduced in this chapter but it could be treated earlier, perhaps when mass is discussed in Chapter 3. Density works nicely here however, because it plays a central role in the chapters that follow. (I like introducing ideas when they are needed, so they aren't considered excess baggage.) Table 18-1 in the text lists the densities of some common materials. Not included is the density of atomic nuclei. Atomic nuclei comprise a tiny fraction of space within matter and have a density of about 2×10^{14} gm/cm^3. When crushed as in the interior of neutron stars, atomic nuclei have a density of about 10^{16} gm/cm^3. That's dense!

Here is something interesting not in the text. Materials denser than lead are very expensive (platinum, gold, mercury), so they are generally not used when denseness is wanted. For instance, a lead weight may not be as dense as a gold one, but lead is more affordable. Uranium, on the other hand, is dense and relatively cheap, and it is used as a substitute for lead in some cases (such a bullets in naval Gatling guns, or low-volume, heavy weights at the bottom of boat keels.)

Students should find text Section 18.5, *Scaling*, particularly interesting. Many more examples on the implications of surface to volume ratio can be explored. For instance, the readings *On Being the Right Size* by J.B.S. Haldane and *On Magnitude* by Sir D'Arcy Wentworth Thompson are good ones. These are both in *The World of Mathematics, Vol. II* edited by J.R. Newman (Simon & Schuster, 1956).

Scaling is becoming enormously important as more devices are being miniaturized. Researchers are finding that when something shrinks enough, whether it is an electronic circuit, motor, film of lubricant, or an individual metal or ceramic crystal, it stops acting like a miniature version of its larger self and starts behaving in new and different ways. Paladium metal, for example, which is normally composed of grains about 1000 nanometers in size, is found to be five times as strong when formed from 5 nanometer grains.

Our technology is changing from "top down" to "bottom up." In the top-down method, relatively large pieces of material are carved into smaller pieces. Such has been the case in the era of milling machines and lathes. In the bottom-up method, matter is assembled an atom at a time, which is nature's way. Trees grow an atom at a time. Will tomorrow's human-made devices in the era of *nanotechnology* do the same?

Suggested Lecture

Begin your lecture by showing evidence for crystal structure in solids. Pass small sheets of galvanized iron around the room.

Density: Pass around two objects of about the same volume but vastly different mass, like a lead block and a wooden block. Have your students shake each one. This should dispel the notion that mass is volume. Go further using the same mass but different volume. Shake the objects vertically, then horizontally. Shaking the object vertically makes weight evident, but weight is not involved with horizontal shaking. This leads to the vital concept that mass density is basic, but weight density is not. Weight density depends on gravity.

An interesting fact to share is that the metals lithium, sodium, and potassium are all less dense than water and will float in water.

Measure the dimensions, in cm, of a large wooden cube and find its mass with a pan balance. Define density as mass divided by volume. Some of your students will unfortunately still confuse density as massiveness or bulkiness, even after they give a verbal definition properly.

CHECK QUESTIONS: When an object is cut into pieces, what happens to the density of each piece? [Each piece has the same density as the original object.] Which has the greater density, a kilogram of lead or a kilogram of feathers? [Any amount of lead is more dense than any amount of feathers.] Which is more dense, a single uranium atom or the world? [The uranium atom]

Acknowledge weight density as it is used in the British system of units, where the density of water is 62.4 lb/ft^2. Discuss the boxed section on text page 255. It is the essence of the "Eureka" story of Archimedes and the gold crown.

Elasticity: Introduce elasticity with an activity version of *Stretch*, which is Experiment 43 in the Lab Manual.

DEMONSTRATION [18-1]: Hang weights from a spring to illustrate Hooke's Law. Ask the class to predict the elongations before suspending additional masses on the thin spring. Next place weights on top of the stiff spring and predict the compressions. Hooke's law holds for both stretching and compression.

DEMONSTRATION [18-2]:   Drop spheres of glass, steel, rubber, and other materials onto a hard surface and compare the elasticity of each material. Your students will be surprised to see that glass and steel are considerably more elastic than rubber is!

Compression and Stretching: Look at text Figures 18-7 and 18-8 and discuss the reason for the I-beam cross section on construction girders. Mention the following fact not in the text: One cubic inch of bone can withstand a 2-ton force.

DEMONSTRATION [18-3]:   Show the weakness of a paper drinking straw when a force is applied perpendicular to its length. A small load buckles the straw. However, along its length the straw has great strength, just like vertical construction girders do. Show this by grasping a straw firmly near one end with your forefinger and thumb and slamming it spear wise against a potato held in your other hand. With very little practice you can pierce the potato with the straw. (Similarly, various reports have cited pieces of straw penetrating boards and other materials during tornadoes. If the straw strikes exactly perpendicular to the material it encounters, penetration rather than bending occurs.)

A fact which is not mentioned in the text and is an interesting follow-through to the text material is the ease with which tiny cracks in strong structures can lead to big breaks. This accounts for disasters such as ships breaking in two or wings falling off airplanes. Bonds between atoms are broken in a crack or scratch. This places greater stress upon the neighboring bonds. A concentration of stress occurs at the tip of a crack, so that a relatively small force can cause the overstrained bond to break. The next bond in turn breaks and the whole structure separates in zipperlike fashion. The initial dislocation need not be a large one. A glass cutter needs only to make a shallow scratch on the surface of the glass, and it will break easily along the line of the scratch. Similarly, a piece of cloth is easily ripped if a small nick is first made in the material.

Area and Volume: Introduce the relationship between area and volume as Chelcie Liu does with the following demonstration.

DEMONSTRATION [18-4]: Have a 500-ml or 1000-ml spherical flask filled with colored water sitting on your lecture table. Produce a tall cylindrical flask of the same volume, but do not tell your students that the volume is the same. Ask them to speculate as to how high the water level will be in the cylindrical flask when all the water from the spherical flask is poured into it. Ask for a show of hands from those who think that the water will reach to more than half the height, from those who think it will fill to less than half the height, and from those who think it will fill to exactly half the height. Your students will be amazed when they see that the seemingly smaller spherical flask has the same volume as the tall cylinder.

To explain the above, call attention to the fact that the area of the spherical flask is considerably smaller than the surface area of the cylinder. The greater area of the cylindrical surface leads one to think that the volume of the cylinder should be greater. Be sure you do this. It is more impressive than it sounds.

Scaling: Get into the most interesting part of the chapter by using the following demonstration to explain text Figures 18-10 and 18-11.

DEMONSTRATION [18-5]:   Display a bunch of cubical toy blocks or a box of sugar cubes. Be sure to show actual cubes and even to pass sets of eight or more cubes to groups of students. Playing with actual cubes will dispel any remnants of confusion between area and volume that might still exist.

Regarding text Figure 18-12, note that the span from ear tip to ear tip of the elephant is almost equal to its height. The dense packing of veins and arteries in the elephant's ears creates a five-degree temperature difference between the blood entering and leaving the ears. A second type of African elephant which resides in cooler, forested regions has smaller ears. Perhaps Indian elephants, which have relatively small ears, evolved in cooler climates.

CHECK QUESTIONS: Which has more surface area, an elephant or a mouse? [Elephant] Which has more surface area, 2000 kilograms of elephant or 2000 kilograms of mice? [2000 kg of mice] Distinguish carefully between these two questions.

CHECK QUESTION: Give two reasons why small cars are more affected by wind. [Small cars have less mass and more cross-sectional area compared to weight than larger cars do.]

CHECK QUESTION: Why is food chopped into small pieces for quick cooking in a wok? [Since cooking occurs from the surface inward, the greater area per volume shortens cooking time.]

CHECK QUESTION: In terms of surface to volume, why should parents take extra care that a baby is kept warm enough in a cold environment? [The baby has more surface per body weight and will cool more rapidly than a larger person will.]

Related facts not mentioned in the text: The surface area of human lungs is about twenty times greater than the surface area of the skin. The digestive tract of an adult human is about 10 m long.

Your lecture can continue by posing exercises from the chapter end-material and by using the check-your-neighbor technique or general class discussion. The examples posed in the Think and Explain questions in the text and the questions in the Lab Manual activity *Geometrical Physics* will perk up class interest.

More Think-and-Explain Questions

1. Is a piece of lead necessarily heavier than a piece of wood? Explain.

   Answer: Although a piece of lead is denser than any piece of wood, it may or may not be heavier, depending on its size. A small piece of lead is obviously lighter than a full grown tree.

2. Why can large fish generally swim faster than small fish can?

   Answer: To understand why large fish swim faster, consider dropping a large fish and a small fish through the air. The larger fish will "plow through" the air more easily, just as a heavier parachutist descends in air faster than a light parachutist does. The situation is the same in water. The large fish has more mass and therefore more strength, compared to its surface area and resistance.

3. Do the effects of scaling help or hinder a small swimmer on a racing team?

   Answer: Scaling effects hinder small swimmers. Large swimmers have more mass per surface area and so more strength compared to the water resistance they encounter.

Computational Problems

1. A one cubic-centimeter cube has sides that are 1 cm in length. What is the length of the sides of a cube with a volume of two cubic-centimeters?

   Answer: Each side is the cube root of 2 cm^3, which is 1.26 cm.

2. Large people at the beach need more suntan lotion than small people do. Compared to a small person, how much lotion will a twice-as-heavy person use?

   Answer: About 1.6 times as much: Because both people have the same density, the person who is twice as heavy has twice the volume, though he has less than twice the surface area. How much more surface area is there for a body with twice the volume? Consider the unit cube of the previous problem. Twice the volume means each side is the cube root of two (1.26) times the side of the smaller cube. Its area is then 1.26 × 1.26, or 1.587 times greater than the smaller cube. Consequently, the twice- as-heavy person at the beach would use about 1.6 times as much suntan lotion.

3. Consider eight, one-cubic-centimeter sugar cubes stacked two by two to form a single large cube. What is the volume of the large cube? How does its surface area compare to the total surface area of the eight separate cubes?

   Answer: The large cube will have the same combined volume of the eight little cubes, but it will have half their combined area. The area of each side of the little cubes is 1 cm^2, and for its six sides, the total area of each little cube is 6 cm^2. Consequently, all eight individual cubes have a total surface area of 48 cm^2. The area of each side of the large cube on the other hand is the square of 2 cm or 4 cm^2. For all six sides the total surface area is 24 cm^2, or half as much as the total surface of the separate small cubes.

4. Consider eight spheres of mercury, each with a diameter of 1 millimeter. When they coalesce to form a single sphere, how big will it be? How does its surface area compare to the total surface area of the previous eight little spheres?

   Answer: The big sphere will have twice the diameter of the 1-mm spheres, and only half as much surface area as the eight little spheres. This problem is similar to the previous one for cubes, since the scaling principles illustrated in the text by cubes hold for any shapes. (The fact that the total surface area is reduced when smaller parts combine to form a larger shape is employed by mice and other creatures who ball up in clusters in cold weather).

19 Liquids

To use this planning guide work from left to right and top to bottom.

Chapter 19 Planning Guide
• *The bulleted items are key: Be sure to do them!*

Topic	Exploration	Concept Development	Application
Liquid Pressure	• Act 45 (>1 period)	• Text 19.1/Lecture Demo 19-1	
Buoyancy		• Text 19.2/Lecture Demo 19-2	Nx-Time Q 19-1
Archimedes' Principle		• Text 19.3/Lecture Demo 19-3 Con Dev Pract Pg 19-1	
Density and Flotation		•Text 19.4, 19.5/Lecture Demo 19-4 Con Dev Pract Pg 19-2	• Exp 46 (>1 period) Nx-Time Q 19-2, 19-3
Pascal's Principle		Text 19.6/Lecture	

Video: *Liquids I, Liquids II*
Evaluation: Chapter 19 Test

See the Chapter Notes for alternative ways to use all these resources.

Objectives

After studying Chapter 19, your students will be able to:
• Describe what determines the pressure of a liquid at any point.
• Explain what causes a buoyant force on an immersed or submerged object.
• Relate the buoyant force on an immersed or submerged object with the weight of the fluid it displaces.
• Describe what determines whether an object will sink or float in a fluid.
• Given the weight of a floating object, determine the weight of the fluid it displaces.
• Describe how Pascal's principle can be applied to increase the force fluid exerts on a surface.

Possible Misconceptions to Correct

• Liquid pressure depends on the total weight of liquid present.

• Immersed and submerged mean the same thing.
• The buoyant force that acts on a submerged object equals the weight of the object, rather than weight of water displaced.
• Heavy things, rather than dense things, sink in water while light things float.
• Whether something sinks or floats depends on its weight, rather than on its density.

Demonstration Equipment

• [19-1]: Pascal's vases
• [19-2]: Overflow can, graduated cylinder or a liquid-measuring cup, and a metal or stone weight to lower into the water-filled vessel by a string
• [19-3] Pair of spring scales, metal block or stone, vessel filled with water, and smaller vessel to catch overflow (as shown in Figure 19-10, text page 272)
• [19-4]: Cartesian diver (Activity 3, text page 281)

Introduction

Knowledge about the ocean and outer space is of current interest, so this and the following chapter should have special interest for your class. Liquids are a real part of your students' everyday world.

Consider doing this neat demonstration. Mix some cornstarch with water in a shallow pan in front of your class. The gooey liquid state is evident as it is sloshed about in the pan. Then invite a student to strike the liquid with a rubber mallet. Instead of splattering, it thuds. Under pressure, the liquid mixture is like putty. It is more a solid than a liquid, much like a pressure-sensitive liquid crystal. For an explanation, see Jearl Walker's delightful book, *The Flying Circus of Physics, with Answers*, (Wiley, 1978).

The unit "ton" is used in several places in this text. It may be taken to mean a metric tonne (the weight of 1000 kilograms) or the British ton (2000 pounds). Either interpretation is sufficient in treating the idea involved.

You may find that many students who have trouble conceptualizing buoyant force are confused about the distinction between area and volume. Be sure to make this distinction clear. If you didn't pour the contents of the 500-ml or 1000-ml spherical flask into the tall same-volume cylinder as described in the suggested lecture of the last chapter, be sure to do it now.

A prerequisite to this chapter is a knowledge of density, which is covered in the previous chapter. So if you skipped Chapter 18, discuss density now. This chapter is a prerequisite for the following chapter, but not for the remaining chapters.

Suggested Lecture

Pressure: Begin by recalling the distinction between force and pressure from Chapter 4. Illustrate with examples. Somebody pushing on your back with a force of only one newton—with a pin! As you're lying on the floor, a woman steps on your stomach—she's wearing spike heels! An Indian master lying on a bed of 1000 nails—the apprentice considering starting with one nail! What was the importance of jewel bearings in watches, a diamond stylus in record players, rounded corners on tables, and sharp blades on cutting knives?

Have students heft a small steel ball and a large plastic foam ball to compare their weights. After they agree that the little ball is heavier (since they know about density from the previous chapter), weigh the balls to show that the plastic ball is heavier! Another example of pressure is the nerve endings of the hand that make one aware of the weight of each ball.

Liquid Pressure: Recall the definition of density from the previous chapter. Consider the pressure at the bottom of a container of water, as is done in text Figure 19-2. Liquid pressure equals weight density × depth. After a few words about weight density, you may want to derive this relationship or call attention to the derivation. (See the footnote on text page 267.) Note that mass density times depth (or height) must be further multiplied by g to give pressure (N/m^2, not kg/m^2). The use of g is avoided when weight density, rather than mass density, is used to develop liquid pressure.

DEMONSTRATION [19-1]: Demonstrate Pascal's vases. Rationalize your results in terms of the supporting forces exerted by the sloping sides of the vases. [For wide-mouth conical vases, conical glass sides push upward on water. Consequently, pressure at the bottom is due in effect only to the weight of the cylindrical part of water above. For the nonvertical part of the narrow-mouth vase, glass pushes down on the water (reaction to water pushing on glass) with just as much force as the weight of a column of water above.]

Ask why dams are built thicker at the bottom. After discussing text Figure 19-3, sketch the top view of several dams on the board and ask which design is best. Next relate this to the ideas of the previous chapter; for example, the shape of stone bridges that actually need no mortar and the arched shape of the tops of windows in old brick buildings. Another illustration is the concave ends of large wine barrels.

CHECK QUESTIONS: Would the pressure be greater when swimming three meters deep in the middle of the ocean than when swimming three meters deep in an ocean tidepool? [No] Would the pressure be greater when swimming three meters deep in the middle of the ocean than when swimming three meters deep in a pond? [Yes, but only because salt water is more dense.]

Compare the densities of fresh and salt water. [Salt water is about 2% denser than fresh water is.]

Explain that blood pressure is taken on the upper arm because it is at the same "depth" as the heart. (See text Think and Explain Question 2). Follow-up information: For the average human, blood takes only about 23 seconds to circulate through the body. Every extra pound of fat you carry requires an extra 200 miles of capillaries (another reason for staying slim!).

Show that buoyancy is a consequence of pressure being depth-dependent. Sketch text Figure 19-6 on the board.

DEMONSTRATION [19-2]: Lower an object on a string into water in an overflow can, a graduated cylinder, or a liquid-measuring cup to show displacement and how easily an object's volume is measured.

Distinguish between immersed and submerged. During a bath, a person is immersed, not submerged, in the water. The difference is in degree of water displacement. Point out that more water is displaced when something is submerged than when it is immersed (partially submerged).

Archimedes' Principle: State how one is buoyed upward when either immersed or submerged in water. Call attention to text Figure 19-11 to show that, because pressure depends on depth, the bottom of a submerged object experiences greater pressure than the top does. This difference in pressures produces a difference in forces, which produces a resulting buoyant force. The amount of buoyant force is given by Archimedes' principle.

DEMONSTRATION [19-3]: Show Archimedes' principle with the apparatus in text Figure 19-10.

Another way to state Archimedes' principle is to cite Newton's third law. If you put something in water that pushes 100 N of water out of the way (displaces 100 N), then the water pushes back with 100 N. The buoyant force is equal and opposite to the weight of the water displaced.

CHECK QUESTION: If an object put into water displaces 500 N of water, how much force does the water exert on the object? [500 N] If you displace 1000 N of water, how much force does the water then exert on you? [1000 N]

Point out that even though a liquid is practically incompressible, its density is not depth dependent. The volume of water decreases by 50 one-millionths of its original volume for every one atmosphere increase in pressure, or equivalently, for each additional 10.3 meters in depth. The density of water near the ocean's surface is practically the same as its density far beneath the surface. Greater variation in density occurs due to temperature difference. Usually a student will inquire about waterlogged objects which lie submerged in water and yet are above the bottom. Such objects are slightly denser than the warmer surface water and are not quite as dense as the cooler water at the bottom. Stress that this is

unusual and objects appreciably denser than water always sink to the bottom, regardless of the depth of the water. Scuba divers do not encounter floating rocks near the bottoms of deep bodies of water!

CHECK QUESTION: Two solid blocks of the same size are submerged in water. One block is lead and the other is aluminum. Upon which is the buoyant force greater? [Both the same]

After discussion, try this one.

CHECK QUESTION: Two solid blocks of identical size are released into the same tank of water. One is made of lead and the other is wood. Which is subject to a greater buoyant force? [Because the wood floats, the buoyant force is greater on the lead since it displaces more water.]

The Role of Density: Discuss the three rules that describe the effect of density on submerged objects (see the bottom of text page 273).

CHECK QUESTIONS: What is the approximate density of a fish? Of a person? What can you say of people who can't float? [The density of a fish equals that of water. Most people are slightly less dense than water. People denser than water can't float.]

It is interesting to note that mountains float on the earth's mantle and are less dense than the molten mantle. Also, just as most of an iceberg is below the water line, most of a mountain is below the surface of the ground. The earth's crust is therefore deeper where there are mountains.

Flotation: Explain how Archimedes' principle holds whether or not an object sinks or floats. In the case of floating, the buoyant force equals not only the weight of the liquid displaced, but also the weight of the floating object.

CHECK QUESTION: Is the above statement a coincidence? [No, the equilibrium of floating requires that the buoyant force be equal and opposite to the gravity force—its weight.]

CHECK QUESTION: What is the buoyant force on a ten-ton ship floating in fresh water? In salt water? In a lake of mercury? [10 tons in each case, although the *volume* of liquid displaced differs]

Discuss boats and rafts and how the water lines change when they are loaded.

Technically speaking, we can say that a completely submerged object that doesn't sink or rise is floating, but we commonly consider floating as being higher than the surface or breaking it. This distinction becomes evident when we consider a balloon floating in the air. Air is a fluid and the same buoyancy principle applies. (This subject is considered in the next chapter).

DEMONSTRATION [19-4]: See the Cartesian diver in Activity 3 on text page 281.

Discuss the compressibility of the human body when swimming. Consider that the density of most people at a meter or two below the surface of the water is still less than the density of water. One need only to relax in order to be buoyed to the surface. However, at greater depths, the increased pressure compresses us to densities that are greater than the density of water, and we must swim to the surface. By simply relaxing, we would sink to the bottom! Relate this to the Cartesian diver demonstration. This is the crux of text Think and Explain Question 7 (the weighted balloon that sinks when pushed beneath the surface).

Pascal's Principle: Begin by pushing against the wall with a meter stick and state that the stick affords a means of applying pressure to the wall. Next state that the same can be done with a confined fluid. Explain how any external pressure applied to a liquid that tightly fills a vessel is transmitted equally to all parts of the liquid . Discuss text Figures 19-18 and 19-19. If a hydraulic press is available, crush a block of wood with it. Point out that the pressure transmitted throughout a confined fluid is pressure over and above that already exerted by the liquid. For example, the pressure in a hydraulic system at any point is equal to the applied pressure plus density times depth.

Point of confusion for some students: Figure 19-19 in the text shows the role of the different areas in multiplying forces. The area of the output piston is larger than the area of the input piston. Figure 19-20 suggests the opposite however, since a large area is shown above the liquid in the reservoir (input) compared to the smaller area of liquid below the piston (output). But input pressure in this case has nothing to do with the surface area in the reservoir. The input pressure is produced by the air compressor and is transmitted to the reservoir and against the output piston in accordance with Pascal's principle.

More Think-and-Explain Questions

1. Next time you're near a farm, notice that a silo has metal bands around it to give it strength.

Notice also that these bands are closer together near ground level and are spaced farther apart near the top of the silo. Why is this so?

Answer: The bands are closer together at the bottom because the pressure of its contents against the sides is greater near the bottom of the silo.

2. Why does water seek its own level?

Answer: Water seeking its own level is a consequence of pressure which is dependent on depth. In a bent U-tube full of water, the water in one side of the tube pushes water up the other side until the pressure in both sides is equal. Consequently, the corresponding depth of water causing these pressures is equal.

3. Suppose you wish to see if a point at the front of your house is at the same elevation as a point around the back. How can you use a garden hose filled with water to determine whether the elevation is equal?

Answer: The use of a water-filled garden hose as an elevation indicator is a practical example of water seeking its own level. The water surface at one end of the hose will be at the same elevation above sea level as the water surface at the other end of the hose. As in the previous question, the hose is simply a widespread U-tube.

Computational Problems

1. The depth of water behind the Hoover Dam in Colorado is 220 m. What is the water pressure at the base of this dam (neglecting the pressure due to the atmosphere)?

   Answer: P = weight density × depth = 9800 N/m^3 × 220 m = 2156 × 10^3 Pa = 2160 kPa (to three significant figures).

2. The top floor of a building is 30 m above the basement. How much greater will the water pressure be in the basement than on the top floor?

   Answer: Water pressure is 294 kPa greater at basement level than on the top floor. P = weight density × depth = 9800 N/m^3 × 30 m = 294 000 N/m^2, or 294 kPa.

3. Which do you think produces more pressure on the ground, an elephant or a woman standing on high heels? Approximate a rough calculation for each.

   Answer: A woman in high heels produces considerably more pressure on the ground than an elephant does! Example: A 400-N woman in 1-cm^2 high heels puts half her weight on each foot and produces a pressure of (200 N/1 cm^2) = 200 N/cm^2. A 20 000-N elephant with 1000 cm^2 feet puts one-fourth its weight on each foot to produce a pressure of (5000 N/1000 cm^2)

= 5 N/cm². This is about 1/40 the pressure. So a woman in high heels will make deeper dents in a linoleum floor than an elephant will.

4. A 0.6-kg piece of metal displaces 1 liter of water when submerged. What is its density?

Answer: Density = m/v = 0.6 kg/1 l = 0.6 kg/l since there are 1000 liters in 1 cubic meter, density may be expressed in units kg/m³. Density = 0.6 kg/1 l × 1000 l/m³ = 600 kg/m³.

5. Calculate the approximate volume of a person with a mass of 100 kg who can just barely float in fresh water.

Answer: 0.1 m³; The person's weight density is about that of water, 9800 N/m³. From the definition density = weight/volume, rearrangement gives volume = weight/density = (100 kg × 9.8 N/kg)/9800 N/m³ = 980 N/9800 N/m³ = 0.1 m³.

6. A gravel barge, rectangular in shape, is 4 m wide and 10 m long. When loaded, it sinks 2 m in the water. What is the weight of gravel in the barge?

Answer: The weight of gravel = weight of displaced water = weight density of water × volume of water displaced = 9800 N/m³ × (4 × 10 × 2)m³ = 784 000 N.

7. Oak is 0.8 times as dense as water and therefore floats in water. What is the weight of the water displaced by a 50-kg floating oak beam? What additional force would be required to poke the oak beneath the surface so that it is completely submerged?

Answer: Since it floats, weight of displaced water equals weight of oak, or mg = 50 kg × 9.8 N/kg = 490 N. Some thought will show that this is 0.8 times the weight of water, *W*, which would be displaced if the oak were completely submerged; that is, 490 N = 0.8 *W*, so *W* = 490 N/0.8 = 612.5 N. The additional force needed to submerge the oak beam is then 612.5 – 490 = 122.5 N. (Like so many problems, there are other ways to arrive at the same result.)

To use this planning guide work from left to right and top to bottom.

Chapter 20 Planning Guide
• The bulleted items are key: Be sure to do them!

Topic	Exploration	Concept Development	Application
Atmospheric Pressure	Act 47 (1 period)	• Text 20.1, 20.2/Lecture Demos 20-1, 20-2, 20-3 Con Dev Pract Pg 20-1	
Barometers		• Text 20.3, 20.4/Lecture Demos 20-4, 20-5	
Boyle's Law		• Text 20.5/Lecture Demo 20-6	• Exp 48 (1 period) Nx-Time Q 20-1
Buoyancy of Air		• Text 20.6/Lecture Con Dev Pract Pg 20-2	Nx-Time Q 20-2
Bernoulli's Principle		• Text 20.7, 20.8/Lecture Demo 20-7	Nx-Time Q 20-3

Video: *Atoms*
Evaluation: Chapter 20 Test

See the Chapter Notes for alternative ways to use all these resources.

Objectives
After studying Chapter 20, students will be able to:
- Explain what prevents the molecules in the earth's atmosphere from either escaping or settling to the ground.
- Describe the source of atmospheric pressure.
- Explain why water cannot be raised higher than 10.3 m with a vacuum pump.
- Describe the relationship between pressure and density for a given amount of a gas at a constant temperature.
- Explain what determines whether an object will float in air.
- Describe the relationship between the speed of a fluid at any point and the pressure at that point, when the flow is steady.
- Explain the principle source of lift on the wing of a bird or airplane.

Possible Misconceptions to Correct
- Air has no weight.
- The atmosphere of the earth extends upward for hundreds of kilometers.
- Things float in air for different reasons than things float in water.
- The faster a fluid moves, the greater is its pressure.
- Atmospheric pressure is greater during a hurricane or tornado.

Demonstration Equipment
- [20-1] A gallon-size metal can and a hot plate
- [20-2] Floppy rubber sheet with a handle in center
- [20-3] Wooden shingle and a sheet of newspaper
- [20-4] Glass of water and two drinking straws
- [20-5] Siphon
- [20-6] A drinking glass and a Ping-Pong ball or other object that will float
- [20-7] Bernoulli examples: beach ball and blower or vacuum cleaner, Ping-Pong ball and hair dryer, cardboard tube with sandpaper inside, Ping-Pong ball on a piece of string and a stream of water

Introduction

The concepts of fluid pressure, buoyancy, and flotation introduced in the previous chapter are applied to the atmosphere in this chapter. Be sure to point out that, unlike a liquid, the density of the atmosphere is depth-dependent. Atmospheric density decreases with increasing altitude.

To avoid information overload, the section on Boyle's law avoids distinguishing between absolute pressure and gauge pressure. An illustration of this distinction is a pressure gauge which registers zero pressure for a flat tire, when in fact the air pressure in a flat tire is atmospheric pressure. Consequently, absolute pressure equals gauge pressure plus atmospheric pressure.

Charles' Law is not covered and reference to temperature effects is made only in a footnote. The plow can be set deeper in a follow-up course.

If you're into lecture demonstrations, this is the material for a show. Two good sources I have found useful are *A Demonstration Handbook for Physics*, by G.D. Frier and F.J. Anderson, published by AAPT, and *Invitations to Science Inquiry*, by Tik L. Liem, 1987.

Blowing bubbles is always fun and can be used to illustrate Bernoulli's Principle. Question: Can you blow a bubble bigger than your lungs in one breath? Answer: Yes, it depends on how you do it. Here's how: Tape together two or three small juice cans that have had both ends removed. (A cardboard core of a roll of paper towels works also, but it will not last through repeated uses.) Make up a soap solution of liquid dish detergent, glycerine, and water. Mix together 1 gallon of water, 2/3 cup of dish detergent, and 3 tablespoons glycerine (available from any drug store.) Let the solution stand overnight, since better bubbles are produced by an "aged" mixture. Dip the tube into the solution to form a soap film over the end. To make a lung-sized bubble, take a deep breath and, with your mouth sealed against the non-soapy end of the tube, exhale and blow a bubble. Don't blow too hard or the bubble film will break. Note the size of this bubble is nearly the volume of your lungs. (You can't exhale all the air from your lungs.)

Now dip the tube into the solution again and this time blow a full breath of air into the tube end, but keep your mouth about 10 cm away from the end. If you are careful, you can blow a bubble as big as your entire upper body! The reason is that when you blow air into the tube, the moving air is at a lower pressure than the stationary air beside it. The stationary air is then drawn into the low-pressure region. (Strictly speaking, air is pushed in by the surrounding atmosphere.) This extra air drawn into (or pushed into) the original stream further inflates the bubble, adding to the amount of air you exhale from your lungs. Very nice!

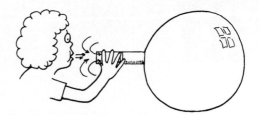

The text credits Bernoulli's principle and the airfoil shape of wings as the explanation of wing lift. But, as the footnote on text page 295 suggests, wings will work without the airfoil. Remember those model planes you flew as a kid, the kind constructed of flat wings? The slot to hold the wing was cut with an "angle of attack." In this way, oncoming air is forced downward. Newton's third law explains the rest. If the wing forces air downward, the air simultaneously forces the wing upward. Birds were able to fly before the time of Daniel Bernoulli, but the question sometimes raised is: Could birds fly before the time of Isaac Newton?

Trying to explain Bernoulli's principle in terms of the differences in molecular impact on the top and bottom surfaces of the wings turns out to be very challenging, especially when experiments show that molecules don't impact the top surface anyway. A thin boundary layer of air is carried in this low pressure region. This boundary layer is evidenced by the dust found on the surface of fan blades!

If in discussing the global atmosphere you address the topic of acid rain and other abuses the atmosphere is undergoing, include a discussion on what can be done to improve the situation. There is no shortage of information regarding the abuses of technology. However, students hear less often how technology can be used to improve the quality of life in the world.

The computer program *Gas* on the computer disk *Good Stuff!* complements this chapter. In this program, 125 gas atoms move about colliding with one another and with the walls of the container. Their speeds are graphed at the side of the picture and you can press "t" to trail one atom along its random path.

This chapter is not a prerequisite for the following chapters.

Suggested Lecture

The Atmosphere and Atmospheric Pressure:
Draw a circle as large as possible on the chalkboard, and then announce that it represents the earth. State that if you were to draw another circle, indicating the thickness of the atmosphere surrounding the earth to scale, you would end up drawing over the same line since over 99% of the atmosphere lies

within the thickness of the chalk line! (We are comparing a 30 km depth of atmosphere to the earth's radius of 6370 km.) Continue your lecture by discussing the ocean of air we live in.

DEMONSTRATION [20-1]: While discussing the preceding topics, have a one-gallon metal can with a small amount of water in it heating on a burner. When it starts to steam, remove the can from the heat source and cap it tightly, as shown in text Figure 20-10. Continue your discussion and the collapsing can will interrupt you as it crunches.

While this is going on, state that if you had a bamboo pole that was 30-kilometer tall with a cross section of one square centimeter, the mass of atmospheric air in it would amount to about one kilogram. The weight of this air is the source of atmospheric pressure. The atmosphere bears down on the earth's surface at sea level with a pressure that corresponds to a weight of one kilogram per square centimeter. That's nearly 10 N/cm^2. Since there are 10 000 cm^2 in 1 m^2, that's 100 000 N/m^2 of pressure. Put it another way. Ask your class to imagine a sewer pipe that is 30-kilometer tall with a cross section of one square meter and is filled with the air of the atmosphere. How much would the enclosed air weigh? The answer is about 10^{5}N. So if you draw a circle of one square meter on the lecture table and ask for the weight of all the air in the atmosphere above, you should elicit a chorus, silent or otherwise, of "10^{5}N!" This is the atmospheric pressure at the bottom of our ocean of air. If your table is located above sea level, in a mountain area for example, then the weight of the atmosphere is correspondingly less (just as water pressure is less for a fish close to the surface of the water).

Next estimate the force of the air pressure that collapsed the metal can, both for a perfect vacuum and for a case where the pressure difference is about one-half atmosphere. Estimate the force of the atmosphere on a person. You can estimate the surface area by approximating different parts of the body on the board, leg by leg and arm by arm.

DEMONSTRATION [20-2]: This great demonstration uses a square sheet of soft rubber with some sort of handle at the center. John McDonald of Boise State University uses a 50-gram mass hanger poked through the center works well. Toss the rubber sheet on any perfectly flat surface. It works best on the top of a lab stool. Picking up the rubber by a corner is an easy task, because air gets under it as it is lifted. However, lifting it at the middle is another story. As the middle of the

sheet is raised, a low-pressure region is formed because air cannot get underneath. The rubber sheet behaves like a suction cup, and the entire stool is lifted when the handle is raised.

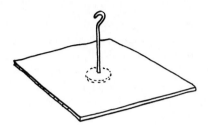

DEMONSTRATION [20-3]: Place a wooden shingle on the lecture table so that it hangs a bit over the edge. Cover the shingle with a flattened sheet of newspaper and strike the overhanging part of the shingle with a stick or your hand (be careful of splinters). Promote more "discuss with your neighbor" activity. [Because parts of the paper are against the table with no air space, the downward force due to the air pressure above the paper is greater than the upward force due to the air pressure below the paper. Consequently, the stick breaks as if someone held the end under the paper down during your blow. Rotational inertia of the stick also contributes to this.]

Barometer: Discuss text Figure 20-8 in which the girl shows that soda cannot be drawn through a straw unless atmospheric pressure (or pressure close to it) acts on the surface of the liquid. Go a step further with the following demonstration.

DEMONSTRATION [20-4]: As the sketch shows, try sucking a liquid through two straws; one in the liquid and the other outside of it. It can't be done since the pressure in your mouth is not reduced due to the second straw. However, with some effort a bit of liquid can be drawn into you mouth. Invite your students to try this and to share this (and other physics ideas!) with their friends.

The vacuum pump, such as the one being demonstrated in text Figure 20-9, operates by atmospheric pressure on the surface of water in the well. When the plunger is lifted, air pressure is reduced in the pipe extending into the well. (The air is "thinned" as more volume is available for the same amount of air.) The greater atmospheric pressure on the surrounding surface of the well water pushes water up the pipe until it finally overflows at the spout.

If all the air could be pumped out of the pipe, the atmosphere would push water up to a height of 10.3 meters. If the well contained liquid mercury, which is 13.6 times more dense than water, the height of water needed to balance the atmosphere would be 1/13.6 of 10.3 meters, or 76 cm. Now discuss the mercury barometer. If you have the opportunity, construct a mercury barometer in front of your class.

CHECK QUESTION: How would the barometer level vary while ascending and descending in the elevator of a tall building? [The level would drop when ascending, and rise when descending.]

Discuss the phenomenon of ear-popping in airplanes and why the cabin pressure in the plane is lower than atmospheric pressure at high altitudes.

DEMONSTRATION [20-5]: The siphon: This water-filled tube with the short end in a reservoir attracts attention, because it is a device where water runs uphill. Water will flow continuously as long as the tube is full and the outlet is below the surface level of the water in the reservoir. The pressure of air on the surface of the reservoir keeps the tube full of water, as in a barometer. That's why the short end of the tube cannot extend upward more than 10.3 m. For a uniform tube, the weight of water in the long end is always greater than the weight of water in the short end. Thus, like a rope dangled over a pulley, the heavy side falls and the light side rises. This difference in weight together with the effect of air pressure in keeping the shorter side filled with water causes the continuous flow. (A more rigorous explanation involves pressures rather than weights of water in each side. Water always flows in an upward direction in the short tube and downward in the long tube, whatever the tube diameter.)

Siphons are commonly used for drawing liquids from large containers into smaller ones. They may be used in bringing water from a pond or spring over higher terrain to any place lower than the source. By means of the *inverted siphon*, water in a pipe may be taken under a river and up on the other side. Water will flow through this pipe without the air of pumps, if the outlet is lower than the place where water enters the pipe.

Boyle's Law: Discuss and/or show Activity 1 on text page 297. This activity involves dunking an upside-down glass in water to show that the "empty" glass contains air and how that air is compressed at greater depths. Relate this to the compressed air breathed by scuba divers. Discuss the reason for the difficulty of snorkeling at a depth of one meter and why it will not work for greater depths. Air will not move by itself from a region of lesser pressure (the air at the surface) to a region of greater pressure (the compressed air in the submerged person's lungs). When we breathe, our diaphragm depresses to reduce lung pressure so the atmospheric pressure outside the body allows a flow of air inward. However, the pressure in the lungs is greater beneath the water due to increased pressure of the water. Any flow of air is outward. That's why pumps must be used to supply air to divers below the surface of water.

DEMONSTRATION [20-6]: Tip a drinking glass upside-down and immerse it in water. A Ping-Pong ball or any floating object in the glass shows that the water level is depressed. This indicates that air is a substance and takes up space.

Relate the air space in the immersed drinking glass to the diving bells used in underwater construction. Workers inside such devices are able to construct bridge foundations and other structures. They are working in an environment of compressed air. How great is the air pressure in these devices? [It is at least as much as the combined pressure of water and the atmosphere outside.]

Buoyancy of Air: Archimedes' principle applies to fluids, gases as well as liquids, discussed in the previous chapter. Objects immersed or submerged in water are buoyed up because there is greater pressure against the bottom of the object than against the top. This occurs simply because the bottom is deeper. The same is true for things buoyed upward by air. Ask the class if the atmospheric pressure is *really* greater at shoulder level than at the top of their heads. [It is, because shoulder level is deeper in the ocean of air than head level is.] Ask for some evidence of this greater pressure and then produce a helium-filled balloon about the size of one's head. The fact that the balloon is visibly buoyed upward is evidence that the atmospheric pressure against the bottom of the balloon (held at shoulder level) is greater than the atmospheric pressure against the top (at head level). Pressure really is depth dependent in both gases and liquids.

CHECK QUESTION: Upon which is the greatest buoyant force, a truck or a child's helium-filled balloon? [The truck, it displaces more air.] Upon which is the buoyant force more effective? [The balloon, the buoyant force is enough to make it rise.] Make the distinction between the thing itself (buoyant force) and the effect of the thing (flotation). Making such distinctions is critical thinking.

Ask if there is a buoyant force on people due to the atmosphere. If there is, why is it not evident? [There is a buoyant force, but it is small compared to one's weight.] Use the following equation to compare the buoyant force on a person to their weight.

BF/weight of person
= weight of air displaced/weight of person
= density of air x volume of air/density of person × volume of person
= density of air/density of person

The density of air is 1.2 kg/m^3 and the density of a person is approximately equal to the density of water, 0.001 kg/m^3, so

BF/weight of person
= density of air/density of person
= 0.0012
= 0.12%

This means that we are buoyed by about one tenth of one percent our weight—that's negligible. If our weight were reduced to one tenth of one percent of our present weight while our volume remained constant, then we would float around like helium-filled balloons.

Bernoulli's Principle: Introduce Bernoulli's principle by blowing across the top surface of a sheet of paper (see text Figure 20-18). Blow extra large "Bernoulli bubbles", which we described earlier. Follow this up with any of the following demonstrations.

DEMONSTRATION [20-7]: (1) Make a beach ball hover in a stream of air from the reverse end of a vacuum cleaner. (2) Do the same with a Ping-Pong ball in the air stream of a hair dryer. (3) Line a cardboard tube with sandpaper and sling the ball in a sidearm fashion. The sandpaper will produce friction to make the ball roll down the tube and emerge spinning. You will see that the ball breaks in the correct direction. Point out that paddles have a rough surface like sandpaper for the same reason—to spin the ball when it is properly struck, that is, to apply "English" to the ball. (4) Swing a Ping-Pong ball taped to a string into a stream of water as shown in text Figure 20-22. Follow this up with a discussion of the shower curtain in the last paragraph of the chapter.

A neat consequence of Bernoulli's principle involves the thickness of a fire fighter's hose with water moving through it when it is at rest. When the hose is turned off but still under pressure, it is fatter than when it is turned on. The faster the water flows in the hose, the thinner it becomes. That's because the pressure in the hose drops as the water moving through it increases in speed.

The curving of pitched balls can be more complicated than the text suggests when different surface textures are taken into account. An excellent resource for more on this is Brancazio's informative book, *Sport Science*.

More Think-and-Explain Questions

1. Compare the density of air at the foot of a deep mine shaft to air density at ground level.

   Answer: The density of air is greater at the bottom of a mine shaft, because the air filling up the shaft is more squashed at the bottom due to its own weight.

2. Why do your ears "pop" when you ascend to higher altitudes?

   Answer: Our inner ear acclimates to the surrounding air and pushes outward just as hard as the atmosphere pushes in against it. At higher altitudes where atmospheric pressure is less, this outward push is greater than the inward push of the atmosphere, hence our ears pop.

3. A balloon filled with air falls to the ground, but a balloon filled with helium rises. Why?

   Answer: The air-filled balloon weighs more than the buoyant force on it, so it falls. The weight of the helium-filled balloon is less than the buoyant force, so it rises.

4. Would a balloon rise in an atmosphere where the pressure was somehow the same at all altitudes? Would a balloon rise in the complete absence of atmospheric pressure, as would occur at the surface of the moon?

   Answer: In both cases, the balloon would not rise, it would fall. This happens because buoyant force is the result of a greater pressure pushing up on an object than is pushing down on it. Without this pressure difference, there would be no buoyant force.

5. What provides the lift to keep a toy disk in flight?

   Answer: Air moves faster over the spinning upper surface of the toy disk, and pressure there is reduced in accordance with Bernoulli's principle. The bowl shape underneath carries along with it a relatively dead air space, which exerts a near-normal pressure upward against the bottom. This difference in pressures against the disk produces upward lift. Usually the lift is nearly equal to the weight and the toy disk remains airborne significantly longer than an ordinary projectile would.

Computational Problems

1. Suppose that air at normal pressure has a mass of 1.30 kg/m³ and hydrogen has a mass of 0.09 kg/m³. How large must the gasbag of a balloon be if the total load has a mass of 300 kg and if the bag is to be filled with hydrogen at normal pressure and at 0°C?

   Answer: 248 m³: By Archimedes' principle, the weight of hydrogen plus the load equals the weight of air displaced. In terms of mass, the mass of the hydrogen plus the load equals the mass of air displaced. The mass of each gas equals density times volume. So we can say:

   Density of hydrogen × volume + 300 kg
   = density of air × volume

   0.09 kg/m³ x volume + 300 kg = 1.30 kg/m³ × volume

   Collect volume on one side of the equation.
   (1.3 - 0.09)kg/m³ × volume = 300 kg
   volume = 300 kg / (1.3 - 0.09)kg/m³ = 300/1.21
   volume = 248 m³.

   Another solution is this: The buoyant force will equal the difference in the weight of air and the weight of an equal volume of hydrogen, and it must also be equal to the weight of 300 kg. In terms of mass, the difference in the mass of air and the mass of hydrogen must equal the mass of the load, which is 300 kg. That is, (1.3 - 0.09)kg/m³ × volume = 300 kg, where volume = 300 kg / (1.3 - 0.09)kg/m³ = 300/1.21 = 248 m³.

2. An automobile is supported by four tires inflated to a gauge pressure of 180 kPa. The area of contact of each tire (ignoring the effects of tread thickness) is 190 cm², which means the total area of tire contact is 0.076 m². Find the mass of the car in kilograms.

   Answer: 1396 kg: From the definition of pressure, 180 kPa pressure equals weight of the car divided by the area of tire contact. By rearranging the equation, we get weight of car = 180 kPa x area of tire contact = 180 × 10³ N/m² × 0.076 m² = 13 680 N. The mass is found from $W = mg$, where m = 13 680 N/9.8 N/kg = 1396 kg. (Note that 1 kPa = 10³ N/m², and g = 9.8 m/s² = 9.8 N/kg.)

3. How many newtons of lift are exerted on an airplane's wings when the difference in the air pressure below and above the wings is 5% of atmospheric pressure? Assume the wings have an area of 100 m³.

   Answer: Lift = ΔP × area = (0.05 × 10⁵ N/m²) × 100 m² = 500 000 N. (This is a mass of about 50 000 kg.)

4. The density of liquid air at standard temperature and pressure is about 900 kg/m³. What will be the volume of 1 m³ of liquid air when it changes to the gaseous state?

   Answer: 692 m²: The volume of 900 kg of air when it reaches a density of 1.30 kg/m² is vol = mass/density = 900 kg/1.30 kg/m² = 692 m².

5. Estimate the ratio of the atmospheric buoyant force per body weight for an average person.

   Answer: 0.0012, or 0.12%: Buoyant force/weight = weight of air displaced/weight of person = density of air x volume/density of person × volume = density of air/density of person (since the volumes are the same). The density of air is about 1.2 kg/m³. The density of an average person is about that of water, 1000 kg/m³. Therefore, buoyant force/weight = density of air/density of person = (1.2 kg/m³)/(1000 kg/m³) = 0.0012, or 0.12%. This means we are buoyed up by about one tenth of one percent of our weight. That's quite negligible.

21 Temperature, Heat, and Expansion

To use this planning guide work from left to right and top to bottom.

Chapter 21 Planning Guide
• *The bulleted items are key: Be sure to do them!*

Topic	Exploration	Concept Development	Application
Temperature and Heat	• Act 49 (1 period)	• Text 21.1, 21.2/Lecture	
Thermal Equilibrium		• Text 21.3/Lecture Con Dev Pract Pg 21-1	
Internal Energy and Quantity of Heat		• Text 21.4, 21.5/Lecture	
Specific Heat Capacity	• Act 50 (1 period)	• Text 21.6, 21.7/Lecture	Exp 51 (>1 period)
Thermal Expansion		• Text 21.8, 21.9/Lecture Demo [21-1] Con Dev Pract Pg 21-2	Exp 52 (> 1 period) Nx-Time Q 21-1, 21-2

Video: *Heat I—Heat, Temperature, and Expansion*
Evaluation: Chapter 21 Test

See the Chapter Notes for alternative ways to use all these resources.

Objectives

After studying Chapter 21, the students will be able to:
• Define temperature and explain how it is measured.
• Describe the relationship between temperature and kinetic energy.
• Define heat and explain why it is incorrect to think of matter as containing heat.
• Describe what determines if heat will flow into or out of a substance.
• Distinguish between internal energy and heat.
• Describe how the quantity of heat that enters or leaves a substance is measured.
• Compare the specific heat capacity of different substances, given the relative amounts of energy required to raise the temperature of a given mass by a given amount.
• Give examples of how the high specific heat capacity of water affects climate.
• Give examples of the expansion of solids as they become warmer.
• Explain the function of a bimetallic coil in a thermostat.
• Compare the thermal expansion of liquids to solids.
• Describe the unusual behavior of water as it is heated from 0°C to 15°C.
• Explain why water at certain temperatures contracts as it becomes warmer.

Possible Misconceptions to Correct

• Heat and temperature are the same.
• Objects contain heat.

Demonstration Equipment

• [21-1] Metal ball and ring apparatus
• [21-2] Bimetallic strip and a flame

Introduction

Like the chapters on properties of matter, the emphasis in the chapters on heat is on water and the atmosphere. No attempt is made to familiarize the student with methods of temperature conversion from one scale to another. The conserved effort is better spent on physics.

In the text, temperature is treated in terms of the kinetic energy per molecule of substances. Strictly speaking, temperature is directly proportional to the kinetic energy per molecule only in the case of ideal gases. However, we take the view that temperature is related to molecular translational kinetic energy in most common substances. Rotational kinetic energy, on the other hand, is only indirectly related to temperature. This is illustrated in a microwave oven where the water molecules are set oscillating with considerable rotational kinetic energy. The food is cooked, not by the rotational kinetic energy of the water, but by the translational kinetic energy it imparts to neighboring food molecules. The food molecules are bounced from the oscillating water molecules like marbles that are set flying in all directions after encountering the spinning blades of a fan. If the food molecules did not interact with the oscillating water molecules, the temperature of the food would not change. Temperature has to do with the translational kinetic energy of molecules.

Quantity of heat is given in terms of calories, a departure from the SI unit, joules. Learning about heat is simplified with the calorie, because score-counting is considerably easier. Students taking life science will use the calorie in their further studies, while students taking physical science will use joules in their follow-up courses.

The definition of the calorie on text page 304 implies that the same amount of heat will be required to change the temperature of water 1 C°, whatever the temperature of the water is. This is an approximation. To be exact, a calorie is defined as the amount of heat required to raise one gram of water from 14° to 15° Celsius.

The term *thermal energy* means different things in different textbooks. To avoid a definition that may clash with further study, the term is simply sidestepped in this book. Instead we discuss *internal energy*, which will be consistent with most advanced textbooks. Heat is defined as the energy (internal energy) that flows by virtue of a temperature difference. The term *heat* implies something happening. Heat, like work, implies a process. Just as work is something we do to a body and not something that a body has, heat is something we do to a body and not something a body has. In both cases, the body acquires and may store *energy*.

The concept of heat flow between temperature differences provides some background to the concept of electron flow (current) between electric potential differences discussed in Chapter 34. Aside from this minor exception, the chapter serves as a prerequisite only for the three following chapters dealing with heat transfer, change of state, and thermodynamics.

Discussions of heat invariably include fire, which is not treated in the text. Fire depends on three things—heat, fuel, and oxygen. Technically, fire is a chemical reaction where a material unites so rapidly with oxygen that flames are produced. Remove any one of the three things required and the fire goes out. Heat can be taken away by cooling, oxygen can be taken away by excluding air, and fuel can be removed to another location.

Suggested Lecture

Temperature: Begin by asking students to explain the difference between a hot cup of coffee and a cold cup of coffee. [The molecules in the hot substance move faster.] Next discuss the measure of this energetic jostling and haphazard bumbling of molecules—temperature. Define temperature as the average kinetic energy, or average KE, of molecules or atoms in a substance.

CHECK QUESTIONS: Which has the higher temperature, a cup of boiling water or a pot of boiling water? Which has more energy? [Both have the same *average* KE per molecule, so the temperatures of both are the same. However, there is more energy in the pot of boiling water, simply because there are more molecules. A pot of water will melt more ice than a cup of boiling water will.]

Thermometers: Describe how the increased jostling of molecules in a warming substance results in its expansion, the basis for the common thermometer. Draw a sketch on the board of an uncalibrated thermometer with a mercury-filled reservoir at the bottom. Then describe how the jostling of energy is transferred from the outer environment to the mercury within the reservoir. If placed in boiling water, the jostling action of the water molecules will be transferred to the mercury, which will expand up the tube. State that one could make a scratch on the glass at this level and label it *100*. Next describe how, if placed in a container of ice water, the mercury will give energy to the cold water and the jostling action will slow down. The mercury will contract and fall to a lower level in the tube. One could again make a scratch on the tube and call this point *zero*. Then, if 99 equally-spaced scratches are made between the two reference points, one would have what used to be called a centigrade thermometer. The prefix "centi-" describes each grade as 1/100 the division between the freezing and boiling points of water. Today, this is called a *Celsius* thermometer. A similar exercise can be done for the Farenheit scale, identifying 32° as the freezing point of water and 212° as the boiling point.

CHECK QUESTION   How would the calibration of the thermometer differ if the glass expanded more than the mercury? [The scale would be upside-down, because the glass reservoir would expand and "open up," allowing more mercury to fill it. The mercury level would fall as the temperature increased.]

CHECK QUESTION: On which of the two temperature scales are the calibrations for degrees spaced the closest, and which is the more accurate when temperature is expressed to the nearest whole number? [On the Fahrenheit scale, there are 180 grades between the freezing and boiling points of water, while there are only 100 grades on the Celsius scale. So, the degree marks are closer together on the Fahrenheit scale. The smaller calibrations are more accurate when temperature readings are expressed to the nearest whole number.]

Acknowledge the Kelvin scale and the absolute zero of temperature, which is the temperature at which molecules can give up no more energy. You will return to the Kelvin scale in Chapter 24, *Thermodynamics*.

Difference between Heat and Temperature: Distinguish between temperature and heat. Define heat as the energy that flows by virtue of temperature differences. Temperature is a measurement expressed in degrees (or kelvins), and heat is measured in units of energy—calories (or joules). Heat is the energy that flows from a higher to a lower temperature, and not the other way around—unless external work is done. When heat is added to a substance, the *internal energy* of the substance increases.

An example that illustrates the difference between heat and temperature is the fireworks sparkler (now illegal in some states). The temperature of the sparks from this device is in the range of 2000°C. Yet they strike the skin with no apparent harm. To say they have a high temperature is to say the sparks have a high ratio of energy per molecule—very high, 2000°C worth. But how many molecules are in a single spark? There are relatively few, so very little energy is

transferred as heat. This is a high-temperature, low-heat situation. By analogy, a tiny drop of boiling water touching your skin transfers very little energy to you, compared to a bucket of boiling water spilled on your skin. The energy per molecule and the total energy involved are different concepts.

CHECK QUESTION: When you strike a match, do you warm up the whole world? [Yes, but not to a noticeable amount: If the match were bigger or the world smaller, perhaps you would notice that the warming that occurs.]

CHECK QUESTION: If you heat half a cup of tea and its temperature rises by 4°C, how much will the temperature rise if you add the same heat to a full cup of tea? [2°C] Does the internal energy in a cup of hot tea increase, decrease, or remain the same when it cools? [It decreases.] Where does the internal energy go when a cup of hot tea cools? [It warms the surroundings. Soon the tea will be cooler and the surroundings warmer. The tea and its surroundings will achieve thermal equilibrium at a common temperature.]

Point out that heat and internal energy are the same form of energy, and whether transferring or at rest, it obeys the law of energy conservation. To say that a body cools is to say that something else warms. Energy may be spread around and diluted, but it never disappears if we know where to look!

CHECK QUESTION: A liter of 30°C water is mixed with a liter of 20°C water. What is the final temperature? [25°C, the internal energy lost by the warmer water is gained by the cooler water.]

Specific Heat Capacity: After distinguishing between calories and degrees, lead into the concept of specific heat capacity by asking your class to consider the difference between touching two different pans that have been placed on hot stove burners for one minute. One is an empty iron skillet and one is a pan containing water. You could safely place your hand in the water, even if it were on the stove for several minutes. Ask which has the higher temperature, the empty skillet or the pan filled with water. (Clearly, it is the empty pan.) Ask which absorbed the greater amount of energy. (The water-filled pan did, if it was on the burner longer.) The water absorbed more energy with a smaller rise in temperature! Physics types have a name for this idea—specific heat capacity.

CHECK QUESTION: Which has the higher specific heat capacity, water or land? [Water does, as evidenced by the greater amount of time required to heat water in the sunshine, as compared to heating land. This fact is a major contributor to weather.]

High Specific Heat Capacity of Water:

Cite some examples of water's high specific heat capacity, such as a hot water bottle on a cold winter night, the cooling system in cars, and the climate in places located near water. With the aid of a large world map, globe, or chalkboard sketch, show the sameness of the latitudes of England and the Hudson Bay and of the latitudes of the French Riviera and Canada. Explain that the Gulf Stream holds heat energy long enough to reach the North Atlantic, because water takes a long time to heat and cool. Once it reaches the North Atlantic, the Gulf Stream cools off. However, according to the conservation of energy, if the water cools, something else has to warm. What is that something? [The air] The cooling water warms the air to create winds, which are westerly at that latitude. Consequently, the warmed air moves over the continent of Europe. Without this phenomenon, Europe would have the same climate as regions of northern Canada do. A similar situation occurs in the United States. The Atlantic Ocean off the coast of the eastern states is considerably warmer than the Pacific Ocean off the coast of Washington, Oregon and California. Yet in winter months the east coast is considerably colder. This has to do with the high specific heat capacity of water and the westerly winds. Air that is warmed by cooling water on the west coast moves landward to bring mild winters to Washington, Oregon, and California. On the east coast, as warmed air moves seaward, it leaves the east coast colder in winter months. In summer months, when the air is warmer than the water, the air cools and the water warms. So summer months on the west coast states are relatively cool, while on the east coast, they are relatively hot. The high specific heat capacity of water serves to moderate climates. The climates of islands, for example, are fairly free of temperature variations. San Francisco, a peninsula that is close to being an island, has the most stable climate of any city in the continental United States.

(This is a good place to break.)

DEMONSTRATION [21-1]:   Consider a metal ball and metal ring. Both are the same temperature, but the ball barely fits through the ring. Ask if the ball would fit in the ring if it were heated. (Students will quickly answer no, because the ball will expand upon being heated.) Then ask the more serious question:

If the ring is heated, will the ball fit through it? In other words, will the hole in the ring get bigger, smaller, or stay the same size? [The size of the heated hole increases.] Many students will correctly reason that the thickness of the ring will increase, but they will then conclude that the hole becomes smaller. You can illustrate the answer by cutting the ring into four quadrants and putting the pieces into a hot oven. They all expand. Next bring them back together to form a ring. This should make it clear that the hole expands when the ring is heated. More simply, the circumference as well as the thickness and every other dimension increases. Also, you can support this demonstration by removing a stuck metal lid from a jar by placing it under hot water. This provokes interest in thermal expansion.

Thermal Expansion—Solids: State that steel lengths expand about one part in 100 000 for each Celsius degree increase in temperature. With this in mind, show a steel rod and ask if anybody would be afraid to stand with their stomach between the end of the rigidly-held steel rod and a wall, while the temperature of the steel rod was increased a few degrees. (This is a safe activity, for the slight expansion of the rod would hardly be noticeable.) Now ask if anyone would stand in the same place with a steel rod a few kilometers in length. (This is different. Although the rate of change is the same, the change in length of the rod could well impale you.) Then discuss the expansion joints on large structures, as shown in text Figures 21-9 and 21-10.

DEMONSTRATION [21-2]:   Place the middle of a bimetallic strip in a flame to show the unequal expansion of different metals and the subsequent bending that occurs.

Point out that different substances expand or contract (length, area, and volume) at their own characteristic rates (coefficients of expansion). Consideration must be given to the different expansion rates of materials that interact; for example, the piston rings when aluminum pistons are enclosed in steel cylinders in a car engine, the rockers on bridges (see text Figure 21-9), and the overflow pipe on a steel tank used for gasoline.

Expansion of Liquids: Elaborate on the overflow pipe on automobile gasoline tanks. Discuss how mercury in a thermometer expands more than the solid glass does, and how water overflows when a pot is filled to the brim.

CHECK QUESTION: How different would a thermometer be if glass expanded more with an increasing temperature than mercury does? [The scale would be upside-down, because the reservoir would enlarge (like the hole enlarged in the heated metal ring) and mercury in the column would tend to fill it up as the temperature increased.]

Expansion of Water: To lead into the idea of water's low density at 4°C, ask the class what the temperature was at the bottom of Lake Michigan on a particular date; New Year's eve in 1905, for example. Ask the same about Lake Tahoe in California for any other date. Continue asking about other lakes until many students are responding "4°C."

CHECK QUESTION: Ask the above question for the bottom of a rain puddle outside the building, and be prepared for some to say "4°C." [4°C was the correct answer for the deep lakes, but it is the wrong answer for puddle. The temperature of a shallow puddle would be the same as the temperature of the surroundings.]

Next go into the explanation given in the book regarding how the microscopic slush forms as the freezing temperature is approached to yield a net expansion below 4°C. I haven't done this, but you might show your class a Galileo-type thermometer. This is a small flask with a narrow glass tube filled with colored water, so changes in temperature would be clearly evident by different levels of water in the narrow tube. Surround the flask with dry ice to chill the water rapidly . The water level drops as the temperature of the water decreases, but the rate slows as it nears 4°C, and then the direction of the water level reverses as cooling continues. This expansion of the water is due to the formation of "microscopic slush." The level of water observed, as a function of time, yields the graph in text Figure 21-14.

The exaggerated volume scale in Figure 21-14 should be emphasized, because it is easy for a student to conclude erroneously that a great change in the volume of water occurs during a relatively small temperature change. Despite the warning on the following page, some students will interpret the volume at 0°C to be that of ice rather than ice water.

CHECK QUESTIONS: Will a sample of 4°C water expand, contract, or remain unchanged in volume when it is heated? [Expand] Will a sample of 4°C water expand, contract, or remain unchanged in volume when it is cooled? [Expand] Why would water, instead of alcohol or mercury, be a poor liquid for a thermometer when temperatures near freezing are to be measured? [The column height would be ambiguous in the 0°C to 8°C range, as shown in text Figure 21-14. It's difficult to distinguish between temperatures on either side of 4°C.]

Discuss text Figures 21-14 and 21-15, and the idea of "microscopic slush."

Ice Formation: Discuss the formation of ice; why it forms at the surface and why it floats. Also explain that deep bodies of water don't freeze over in winter because all the water in the lake has to be cooled to 4°C before colder water will remain at the surface and cool to the freezing temperature, 0°C. State that to cool a teaspoonful of water in the lake to 3°C, let alone 0°C, all the water beneath must be cooled to 4°C. Winters are neither cold enough nor long enough for this to happen in the United States.

NEXT-TIME QUESTION: Ask your students to place an ice cube in a glass of ice water at home and to compare the water level at the side of the glass before and after the ice melts. Ask them to account for the volume of ice that extends above the water line after it melts. [The level remains unchanged.] This can be explained from the principles learned in Chapter 19. The floating ice cube displaces its own weight of water. So if the cube weighs a newton, then when placed in the glass, one newton of water is displaced and the water level rises. If it is first melted and then poured in the glass, the water line will again be higher by the same amount, one newton. It's more interesting to account for the volume of floating ice that extends above the water line. The ice expands when freezing because of the hexagonal open structures of the ice crystals. Ask the class if they have any idea of how much volume all those billions and billions of open spaces constitute. That volume equals the volume of ice extending above the water line! When the ice melts, the part above the water line fills in the open structures as they collapse. Discuss this idea in terms of icebergs and how the coastline would change if all the floating icebergs in the world melted. The oceans would rise a bit, but only because icebergs are composed of fresh water. (They form above sea level, break off, and then fall into sea.) The slight rise is more easily understood by exaggerating the circumstances—think of ice cubes floating in mercury. When they melt, the depth of fluid (water on mercury) is higher than before.

More Think-and-Explain Questions

1. Which has the greater amount of internal energy, a giant iceberg or a hot cup of tea?

   Answer: Although the tea has a higher temperature, which is to say its molecules move faster (higher average KE), the iceberg has more internal energy since it has a far greater number of molecules. It's like 1000 people each with 10 cents in their pocket having more money than one person with 10 dollars.

2. Explain what is meant by saying a thermometer measures its own temperature.

   Answer: Objects in the same locality will ultimately come to the same temperature when hot things cool and cool things warm. This is what a thermometer does. Thermometers come to thermal equilibrium with their surroundings and show this on their scale.

3. Would a bimetallic strip work if the two different metals happened to have the same rates of expansion? Is it important that they expand differently?

   Answer: A bimetallic strip will bend when subjected to a change in temperature only if one part changes differently than the other part does. So it is required that the metals expand (or contract) differently.

4. Why is it important to protect water pipes so they don't freeze?

   Answer: It is important to keep water in pipes from freezing, because water expands more than the pipe material does and the pipe may crack when the water in it freezes.

5. If water had a lower specific heat, would ponds and lakes be more likely or less likely to freeze?

   Answer: More likely, this is because the temperature would decrease more as water loses energy, and the water would be more readily cooled to the freezing point.

Computational Problems

1. What would be the final temperature of the mixture of 50 g of 20°C water and 50 g of 40°C water?

   Answer: 100 g of 30°C water

2. What would be the final temperature when 100 g of 25°C water is mixed with 75 g of 40°C water? (The specific heat of water is 1.0 cal/g C°.)

   Answer: Heat gained by cool water = heat lost by warm water.

   $$m_1 c \Delta T_1 = m_2 c \Delta T_2$$
   $$(100) c (T - 25) = (75) c (40 - T)$$

   (Note that common sense dictates that ΔT_1 is final temperature T minus 25° because T will be greater than 25°, and ΔT_2 is 40° minus T because T will be less than 40°. ΔT_1 does not equal ΔT_2 as in Problem 1 above because of the different masses of cool and warm water.) Solving the number work,

   $$100T - 2500 = 3000 - 75T$$
   $$175T = 5500$$
   $$T = 5500/175 = 31.4°C$$

3. What will be the final temperature of 100 g of 20°C water when 100 g of 40° iron nails are submerged in it? (The specific heat of iron is 0.12 cal/gC°.)

   Answer: Heat gained by water = heat lost by nails

   $$(m c \Delta T)_{water} = (m c \Delta T)_{nails}$$
   $$(100)(1) (T - 20) = (40)(0.12)(40 - T)$$
   $$100T - 2000 = 192 - 4.8T$$
   $$104.8T = 2192$$
   $$T = 2192/104.8 = 20.9°C$$

4. Consider an iron rod that is 1 m long expands 1/4 cm when heated. By how much will a 100-m rod of the same material expand when similarly heated?

   Answer: 100 times as much, or 25 cm: The heated rod will be 100.25 m long.

5. Some steel expands 1 part in 100 000 for each Celsius degree increase in temperature. Suppose the 1.3-km main span of the Golden Gate Bridge had no expansion joints. How much longer would the bridge be for an increase in temperature of 10 C°?

   Answer: For a 10°C increase, the steel bridge will expand 10 parts in 10^5, or one part in 10 000. The bridge will expand by one ten-thousandth of 1.3 kilometers, or one ten-thousandths of 1300 meters, which is 13 centimeters. So the main span of the Golden Gate bridge expands 13 centimeters when the temperature increases by 10°C. By formula, $\Delta L = \alpha L \Delta T = (10^{-5})(1\,300\text{ m})(10) = 0.13\text{m} = 13\text{ cm}$.

22 Transmission of Heat

To use this planning guide work from left to right and top to bottom.

Chapter 22 Planning Guide

• *The bulleted items are key: Be sure to do them!*

Topic	Exploration	Concept Development	Application
Conduction		• Text 22.1/Lecture Demos 22-1, 22-2, 22-3	Nx-Time Q 22-1
Convection	Act 53 (1 period) Act 54 (1 period)	• Text 22.2/Lecture Con Dev Pract Pg 22-1	Nx-Time Q 22-2
Radiation	• Act 56 (1 period)	• Text 22.3/Lecture Demo 22-4	
Absorption and Emission		• Text 22.4, 22.5/Lecture Demo 22-5	Exp 55 (1 period) Nx-Time Q 22-3
Newton's Law of Cooling		• Text 22.6/Lecture	
Greenhouse Effect		• Text 22.7/Lecture	

Video: *Heat II—Heat Transfer, Heat II— Heat Radiation*
Evaluation: Chapter 22 Test

See the Chapter Notes for alternative ways to use all these resources.

Objectives

After studying Chapter 22, students will be able to:
• Explain why two materials at the same temperature may not feel like they are the same temperature when touched.
• Explain why porous materials with air spaces are better insulators than nonporous materials are.
• Explain how heat can be transferred quickly through liquids and gases even though they are poor conductors.
• Distinguish between conduction and convection from an atomic point of view.
• Explain how heat is transmitted through empty space.
• Given the color and shininess of two objects, predict which one will absorb radiant energy more easily.
• Compare the ability of an object to emit radiant energy with its ability to absorb it.
• Relate the temperature difference between an object and its surroundings to the rate at which cooling occurs.
• Describe the processes of absorption and emission of the sun's radiant energy by the atmosphere and surface.
• Describe the greenhouse effect on the earth.

Possible Misconceptions to Correct

• Surfaces that feel cooler than others must have a lower temperature.
• A blanket is a source of heat energy.
• Walking barefoot without harm on red hot coals involves only non-physics considerations.
• Only hot things radiate energy.
• The greenhouse effect on the planet earth is undesirable.
• The greenhouse effect is the principal factor in the warming of the florist's greenhouse.

Demonstration Equipment

- [22-1] Metal bar, sheet of paper, and a flame source
- [22-2] A flame source and a paper cup filled with water
- [22-3] Test tube filled with water, wedged ice, and a flame source
- [22-4] Box with hole and painted white on inside (See text Figure 22-14)
- [22-5] Silvered and black container of hot or cold water and thermometers

Introduction

This chapter begins with the conduction, convection, and radiation of heat with the emphasis on bodies of water and the atmosphere. The section on radiation provides some background for electromagnetic waves, which are covered in Chapter 37.

One task of physics teachers is to help students distinguish between science and pseudoscience. One popular misconception is exploited by people who claim special powers and will teach people how to gain these special powers for a fee. The misconception is that it is not possible to walk barefoot on the red-hot coals of burning wood without being harmed. People are led to believe that there is no scientific explanation for this phenomenon. The truth is that many people without any special powers have walked harmlessly on red-hot wooden coals with bare feet. The explanation is an extension of the statement on page 299 of the text. Conductivity as well as temperature must be considered. For example, you can momentarily put your hand in a very hot oven without harm, not because the temperature is low, but because air is a poor conductor of heat. It is common knowledge that wood has low heat conductivity. This is why wood is used for handles on cooking utensils. You'd burn yourself if you reached into a hot oven with your bare hand and grabbed a frying pan with an iron handle. However, you'd be okay if you quickly grabbed one with a wooden handle. Wood is a poor conductor, even when it's hot; it is still a poor conductor when it's red hot. After the surface of a red hot coal of low-conductivity wood gives up its heat, more than a second passes before the appreciable internal energy from the inside reheats the surface. So, although the coal has a very high temperature, it gives up very little heat when brief contact occurs with a cooler surface. For the "believers," mind-over-matter rather than low conductivity is the accepted explanation. A walk over red-hot pieces of iron would be a horrible confrontation with physical reality.

Another popular misconception is that energy is saved if a heater is not completely turned off in an empty house on a cold day. Turning the thermostat down or off is the subject of Think and

Explain text question 9. To illustrate the answer, make up the apparatus shown. This consists of a main reservoir that feeds "heat" into two identical "houses," which leak heat to the environment. The amount of leakage from each is caught by the bottom jars and it can be compared at a glance. Arrange the input flow rates so that equilibrium is established when the "houses" are nearly full of water. Then, input equals outflow. Next turn one input off altogether. After some time, turn it back on until it fills to the level of the other "house." Now compare the differences in the amount of leaked water! Make more comparisons by reducing the inflow, turning it down partway instead of turning it off. This roughly approximates Newton's law of cooling. The leakage rate is highest when ΔT, or in this case ΔP, is greatest.

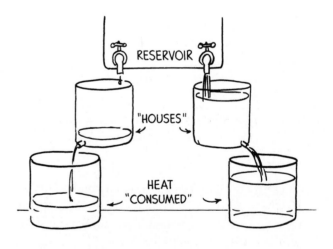

Much of Unit 3 deals with heat as it relates to climate, so an optional lecture on the earth's seasons is included. This topic is not covered in the textbook, but it may follow nicely from your treatment of radiation.

Throughout this manual, I stress the importance of the "check with your neighbor" technique of teaching. Please do not spend your lecture talking to yourself in front of your class! The "check with your neighbor" routine keeps you and your class together. If nothing else, it gives you a chance to look over your notes on the spot, reflect on your presentation, and pace your delivery. For effective teaching, I can't stress its importance enough!

Suggested Lectures

Conduction: Begin by asking why pots and pans have wooden or plastic handles. Then discuss conduction from an atomic point of view, citing the role of the electrons in both heat and electrical conduction. You might demonstrate how wax melts on rods of different metals. Keep the rods equidistant from a hot flame to illustrate the relative conductivity of each metal.

> CHECK QUESTION: How is tipping a row of standing dominoes similar to heat conduction? [In both cases energy is transmitted from one place to another by the collisions of moving units.]

> DEMONSTRATION: [22-1]: Do text Activity 3 by placing a metal bar wrapped in paper in the flame of a Bunsen burner. The paper doesn't catch fire. Why? [The metal conducts heat from the flame so efficiently that the paper cannot reach its igniting temperature of 233°C.]

Other materials can be compared in their ability to conduct heat. Use the example of North American natives preferring their old bone and ivory saws to steel ones for cutting through blocks of snow in igloo construction. Because of the good thermal conductivity of steel, the saw freezes and sticks to the snow, whereas the bone or ivory saws, with their low conductivity, don't stick.

> DEMONSTRATION [22-2]: Place a paper cup filled with water in the flame of a Bunsen burner. The paper will not reach its igniting temperature and burn, because heat from the flame is conducted into the water. Water is not *that* poor a conductor—its high specific heat capacity comes into play here also.

Discuss the poor conductivity of water, which eludes the previous lecture in which you discussed the 4°C water temperature at the bottom of deep lakes all year round.

> DEMONSTRATION [22-3]: Do text Activity 1 (also text Figure 22-5) with the ice wedged at the bottom of a test tube. Some steel wool will hold the ice at the bottom of the tube. It is impressive to see that the water at the top is brought to a boil while the ice below barely melts! (Convection, or the lack of it, is illustrated here as well. If heating occurred at the bottom and the ice cube was at the top, the ice would melt quickly.)

Discuss the poor conductivity of air and its role in insulating materials, such as down-filled sleeping bags and sportswear, spun glass and plastic foam insulation, fluffy blankets, and snow. Discuss thermal underwear and how the fish-net open spaces actually trap air between the skin and the undergarment. This trapped air accounts for the insulating properties. In a similar manner, the air trapped in animal fur accounts for the insulation of fur. While the skin is warm, the temperature at the outer layer of the animal's pelt is about the same as the surroundings. Longer fur is therefore a better insulator.

The fur of a polar bear is more than an excellent insulator. A polar bear's hairs are completely transparent and appear white, because visible light reflects from the rough inner surface of each hollow hair. Ultraviolet light, however, is absorbed and sent to the skin like light traveling in an optical fiber. The bear does not reradiate energy in the UV, so the hairs transmit energy primarily in only one direction—from the outside to the bear and not the other way around (sort of a thermal diode!). The high absorption by the skin is evidenced by its black color! So a polar bear is white on the outside, but black underneath its fur. The polar bear loses very little heat and is almost a perfect solar converter.

Discuss double-pane windows. Assuming you covered atmospheric pressure earlier (Chapter 20), cite the case of the midwestern manufacturer who sent a shipment of double-pane windows by truck over the Rocky Mountains. Upon reaching a higher altitude, he found that all the windows broke. The atmospheric pressure sealed between the panes was not matched by the atmospheric pressure outside. Ask if the windows "imploded" or "exploded." [Exploded, the pressure inside was greater than the atmospheric pressure outside.] Point of information: Double-pane windows do not have vacuums between the panes of glass. They would be squashed together by atmospheric pressure. It is important that no water vapor is in the air space, since condensation can cloud the glass.

Convection: Illustrate convection by holding your fingers beside a flame as shown on text page 320. Ask why you cannot hold your fingers above the flame without being burned.

The lack of convection in orbiting vehicles, such as the space shuttle, has interesting consequences. When in orbit, one cannot light a match without having it snuff out very quickly. One cannot exercise without becoming overheated very quickly. These phenomena are due to absence of convection in orbit. Much of the convection in fluids depends on buoyancy, which in turn depends on gravity. In outer space the local effects of gravity are not there, because the shuttle and everything in the shuttle is in free fall around the earth. So with no convection, hot gases are not buoyed upward away from the flame. They remain around the flame, preventing the entry of needed oxygen and the flame burns out. Without convection to carry heated air upward and away from the body, it quickly overheats. That's why astronauts use fans when they exercise.

Why Warm Air Rises: When a portion of air is heated, it expands and becomes less dense than the surrounding air. Then buoyancy is greater than its weight and the air rises. When it rises, what happens to the surrounding air pressure? [It decreases.] When the surrounding pressure decreases, what happens to the volume of rising air? [It expands.] What happens to its temperature when it expands? [It cools.] Putting it all together, what happens to the temperature of rising warm air? [It cools!]

To see that expanding air cools, have your students blow on their hands. First blow with the mouth open so the air is warm, then with the lips puckered so that the air expands as it leaves the mouth. Noticeable cooling is the result (see text Figure 22-8).

CHECK QUESTION: We all know that warm air rises. So why are mountain tops cold and snow covered, while the valleys below relatively warm and green? Shouldn't it be the other way around? [No, nature is correct — as warm air rises, it cools. In fact, the cool tops of mountains are a consequence of rising warm air. There is no contradiction!]

Discuss the role of convection in climates. Begin by calling attention to the shift in winds as shown in text Figure 22-7. This leads you into radiation and the heat from the sun.

CHECK QUESTION: Why do coastal winds change in direction from day to night? [Land warms faster than water. In the daytime, the land and air above it is warmer than the water and air above it. So the warm air rises and causes a sea breeze from water to land. At night, the reverse happens.]

Radiation: Make a distinction between radiation as it pertains to heat transmission and the radiation of radioactivity. This usage of the same word for two different phenomena is confusing.

Discuss the radiation one feels from red hot coals in a fireplace and how the intensity of the radiation decreases with distance. Consider the radiation one feels when stepping from the shade to the sunshine. Amazing! The heat is not so much because of the sun's temperature (similar

temperatures are found in some welder's torches) but because of its size. Comfortably big!

You may want to discuss why the earth is warmer at the equator than it is at the poles. Get into the idea of solar energy per unit area (as contrasted to the notion that the equator is warmer because it is closer to the sun). A neat way to do this is to ask the class to compare the rays of sunlight striking the earth with vertically-falling rain that strikes two pieces of paper—one held horizontally and the other held at an angle in the rain as shown in the sketch. You can easily dispel the misconception that the paper held horizontally must get wetter than the paper held at an angle because it is closer to the clouds!

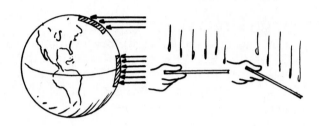

Seasons: This discussion is optional. Depending on the time available, you might consider departing from the text and continuing with the discussion on solar energy per unit of area idea as it relates to the seasons. The plane of the earth's equator is not parallel to the plane of the earth's orbit. Instead, the polar axis is inclined at 23.5 degrees (the ecliptic). Draw the sketch below on the board, first with only the two positions of the earth at the far left and far right. Ask which of these two positions represents winter months and which one represents summer months. Encourage neighbor discussion.

Once it is clear that winter is at the left, show the position of the earth in autumn and in spring. Shift the position of the sun closer to the earth in winter, since this is actually the case. From your drawing, your class can see why Northern-Hemisphere types enjoy an extra week of spring and summer! Southern-Hemisphere types are compensated by a somewhat milder climate year round, due to the greater amount of ocean in the Southern Hemisphere (80% as compared to about 60% for the Northern Hemisphere).

110

Absorption and Emission of Radiation:
There are various colors of eyes, but all of them have one thing in common. The pupils are black. Discuss why this is so.

DEMONSTRATION [22-4]: Make up and show the black hole in the white box, as shown by Helen Yan in text Figure 22-14.

Acknowledge that all things emit radiation, that is all things that have a temperature. But things do not become progressively cooler since they also absorb radiation. We live in a sea of radiation, everything emitting it and everything absorbing it. When the emission rate equals the absorption rate, temperature remains constant. Some materials, because of their molecular design, emit radiation better than others do. They also absorb it better. The good absorbers are easy to spot, because they absorb visible radiation as well and appear black. (Make this distinction; objects don't absorb because they're black, but it is because they absorb so well that they are black. Cause precedes effect.)

DEMONSTRATION [22-5]: If your students did not do Lab Manual Activity 55, *Cooling Off*, check the temperatures of two pairs of black and silvered containers as your class period progresses. Have one pair filled with hot water, and the other pair filled with cold water. (This is much more instructive with two sets than with one set, since you demonstrate both emission and absorption.)

Newton's Law of Cooling: Relate the rate of cooling to the black and silver containers that are cooling and warming. We see the difference between a proportionality sign and an equals sign for the formula here, because the rate of cooling or warming is not only proportional to the difference in temperatures, but also to the "emissivities" of the surfaces. Relate Newton's law of cooling to Think and Explain Questions 7 (cream in the coffee), 8 (cooling a beverage in the refrigerator), and 9 (thermostat on a cold day). These questions make excellent discussion topics.

CHECK QUESTION: Does Newton's law of cooling apply to the warming of a cold object in a warm environment? [Yes]

Greenhouse Effect: Discuss the greenhouse effect, first for florist's greenhouses, and then for the earth's atmosphere. The key idea is that the medium, glass for the greenhouse and the atmosphere for the earth, is transparent to high-frequency (visible) electromagnetic waves, but it is opaque to low-frequency (infrared) electromagnetic waves.

If you wish to go deeper than the text coverage of the greenhouse effect, briefly discuss the idea of wave frequency. Text Figure 22-12 shows the relationship of wave frequency and wavelength. A vigorous shaking of a stretched rope produces high-frequency waves (or short wavelengths). The origin of electromagnetic waves is vibrating electrons in matter. The more frequently they vibrate, the higher is the frequency of waves they emit. This relates to temperature. The equation form, $f \sim T$, reads: The frequency of electromagnetic radiation emitted by a source is proportional to the temperature of the source. Electrons vibrate at greater frequencies in hot matter than in cold matter. The sun is so hot that the frequency of electromagnetic waves it emits are high—high enough to activate our visible receptors. Write $f \sim T$ in big letters on the board to indicate large values of both frequency and temperature. This radiation is visible light. It is absorbed by the earth, which in turn emits its own radiation. Next write $f \sim T$ in small letters to indicate low values of both frequency and temperature. Interesting point: The atmosphere of the earth is primarily warmed by terrestrial radiation, not solar radiation. That's why air near the ground is warmer than the air above it. The opposite would be the case if the sun were the primary warmer of air!

Before linking these ideas to the greenhouse effect of the earth, first consider what happens to the interior temperature of a car on a hot, sunny day. Window glass is transparent to the high frequencies of sunlight, so visible light and the energy it carries comes through the windshield and is absorbed by the car's interior. The car's interior in turn radiates its own electromagnetic waves, but in accordance with $f \sim T$, at a substantially lower frequency than sunlight. These are infrared waves to which the window glass is opaque. These waves can't get through the glass. Hence the glass acts as a one-way valve; high-frequency waves can come in, but low-frequency waves can't get out. Internal energy of the car interior increases. (The warming of the interior of a car, like the heating of the florist's greenhouse, is only partly due to the greenhouse effect. A greater contributor is the glass that traps the warm air, preventing its loss through convection.)

With the car model of the greenhouse understood, go on to the greenhouse effect of the earth. Like the car window, our atmosphere is transparent to visible light but opaque to infrared. This opacity is primarily due to water vapor and carbon dioxide in the atmosphere. The greenhouse effect increases the internal energy of the earth. It turns out that this is a quite desirable effect. Without the greenhouse effect we wouldn't be here, for the average temperature of the earth would be a chilling -18°C! So the controversy over the increased CO_2 in the atmosphere and its effect on the greenhouse effect has to do with the degree of the effect, not the effect itself. Interestingly enough, the carbon that is spewed by burning is the same carbon that is absorbed by tree growth. So a realistic step in the solution to the increased greenhouse effect is to simply grow more trees (while decreasing the rate at which they are cut down)! Johnny-Appleseed types to the task! This would not be an end-all to the problem, however, because the carbon returns to the biosphere when the trees ultimately decay.

Interesting point: The earth is always "in equilibrium" whether it is overheating or not. At a higher temperature that the greenhouse effect produces, the earth simply radiates more terrestrial radiation. Income and outgo match in any case. The important consideration is the temperature at which this income and outgo match.

Solar power has been with humans from the beginning. We see its application whenever we see clothes hung on a line or as a power source on the roofs of new buildings under construction. If you have up-to-date information on this technology, share it with your class here.

More Think-and-Explain Questions

1. Wood is a better insulator than glass. Yet fiberglass is commonly used as an insulator in wooden buildings. Why?

   Answer: Air is an excellent insulator. The reason that fiberglass is a good insulator is principally because of the vast amount of air spaces trapped in it.

2. You can bring water in a paper cup to a boil by placing it over a flame. Why doesn't the paper cup burn?

   Answer: Much of the energy of the flame is readily conducted through the paper to the water. The paper cup and the water comprise a thermal system that, as a whole, will increase in temperature. The relatively large amount of water, compared to the amount of paper, absorbs energy that would otherwise heat the paper. This keeps the temperature of both the cup and the water well below the igniting temperature of paper.

3. A bowl of soup may be too hot to consume comfortably. If the top surface is removed, that layer will very likely be consumable. Why?

   Answer: The hottest part of the soup is the least dense and floats at the surface.

4. On sunny days, why do hot-air balloons suddenly rise when they drift over a wide road or parking lot made of black asphalt?

   Answer: The air warmed by the black surface is rising.

5. Why does a good emitter of heat radiation appear black at room temperature?

   Answer: The reason that a good absorber appears black is discussed in the footnoted answers on text page 325. A good absorber of radiation is, by design, also a good emitter. This is evident when a good absorber is not generally any warmer than poor absorbers are in the same environment. This balance is called thermal equilibrium. A good emitter appears black at room temperatures, because the radiation it is emitting is too low in frequency to be seen. The rest is logic. Since a good absorber appears black and since a good absorber is also a good emitter, it follows that a good emitter appears black. (Heat a normally black body to incandescence, and you'll find it is a better emitter than a non-black body of the same temperature.)

6. Why are space shuttle heat-insulating tiles black?

   Answer: They are excellent absorbers of heat, as evidenced by their blackness.

7. A number of objects at different temperatures placed in a closed room will ultimately come to the same temperature. Would this thermal equilibrium be possible if good absorbers were poor emitters and if poor absorbers were good emitters? Explain.

   Answer: If good absorbers were not good emitters, thermal equilibrium would not be possible. If a good absorber only absorbed, then its temperature would climb above that of the poorer absorbers in the vicinity. Also, if poor absorbers were good emitters, their temperatures would fall below that of the better absorbers.

23 Change of State

To use this planning guide work from left to right and top to bottom.

Chapter 23 Planning Guide

• *The bulleted items are key: Be sure to do them!*

Topic	Exploration	Concept Development	Application
Evaporation	Act 62 (1 period)	• Text 23.1/Lecture	
Condensation		• Text 23.2, 23.3/Lecture Demo 23-1	
Boiling	Act 57 (1 period)	• Text 23.4/Lecture Demo 23-2	
Freezing		• Text 23.5, 23.6, 23.7/Lecture Demos 23-3, 23-4	Exp 58 (>1 period)
Energy Changes	Act 60 (>1 period)	• Text 23.8/Lecture Con Dev Pract Pg 23-1, 23-2	Exp 59 (>1 period) Exp 61 (1 period) Nx-Time Q 23-1, 23-2

Video: *Heat IV—Change of State*
Evaluation: Chapter 23 Test

See the Chapter Notes for alternative ways to use all these resources.

Objectives

After studying chapter 23, students will be able to:

- Explain why evaporation is a cooling process.
- Explain why condensation is a warming process.
- Explain why a person with wet skin feels chillier in dry air than in moist air at the same temperature.
- Distinguish between evaporation and boiling.
- Explain why food takes longer to cook in boiling water at high altitude than it does at sea level.
- Explain why water containing dissolved substances freezes at a lower temperature than pure water does.
- Describe the circumstances under which something can boil and freeze at the same time.
- Give examples of the tendency of ice to melt under pressure and refreeze when the pressure is removed.
- Describe the conditions needed for a substance to absorb or release energy with no resulting change in temperature.

Possible Misconceptions to Correct

- Constant temperature of something indicates that all the molecules have the same energy.
- Boiling is a warming process.
- Ice melts only when heat is added.

Demonstration Equipment

- [23-1] Aluminum soda pop cans, hot plate, pan of water
- [23-2] Flask of water with enough air pumped out so water boils by the heat of one's hand
- [23-3] Triple-point apparatus (text Figure 23-10)
- [23-4] Ice, copper wire, and weights (text Figure 23-11)

Introduction

In this chapter, the emphasis is again on bodies of water and the atmosphere. Material from this chapter is not a prerequisite for chapters that follow.

Note that the calorie unit is used to express the heat of fusion and vaporization of water. The SI units have their merits, but they have their drawbacks too. I have a strong bias for saying that one calorie will raise the temperature of one gram of water by 1°C, rather than saying 4186 J will raise the temperature of 1 kg of water by 1°C, and for saying 80 calories will melt one gram of ice and 540 calories will vaporize one gram of boiling water, rather than using the SI figures 334.88 kJ/kg and 2260 kJ/kg. I find the SI values somewhat more difficult conceptually. If you're a 100% SI type, the footnotes on text page 338 give the SI units. You can lecture with SI units and point out the few places where the unit calorie occurs.

If you wish to introduce the idea of distribution curves into your course, this is a good place to do it. Treat the cooling produced by evaporation by plotting on a graph the number of molecules in a liquid versus the molecule's speed. Show how the distribution shifts as the faster-moving molecules evaporate. You may wish to point out that the bell-shaped distribution curves are used to represent the distribution of many things, from molecular speeds to examination scores. Regrettably, many people tend to regard such distributions not as bell-shaped, but rather as spikes. This makes a difference in attitudes. For example, suppose you compare the grade distributions for two sections of your course, Group 1 and Group 2, and the average score for Group 1 is somewhat greater than it is for Group 2. For whatever reason, Group 1 outperforms Group 2. With this information can we make any judgement about an individual from either group? One who looks at these distributions as spiked shapes believes he can. He will say (or think) that all individuals from Group 1 do better than any individual from Group 2. On the other hand, one who thinks in terms of the breadth of the bell-shaped distribution will not make any assumptions about each individual. We should all be aware of the region of overlap in the two distribution curves. Attitudes toward individuals from either group should be unbiased by unwarranted prejudice. Hence the difference between being narrow-minded and broad-minded!

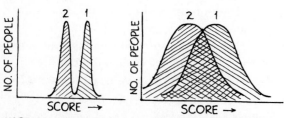

NARROW-MINDED PERCEPTION   BROAD-MINDED PERCEPTION

Suggested Lecture

Evaporation: Begin by asking: What does it feel like to leave the water after swimming? (The air is very chilly, especially when it is windy.) Explain the cooling of a liquid from an atomic point of view. Also reinforce the idea of temperature being a measure of the average molecular kinetic energy, which means there are molecules that move faster and slower than the average.

CHECK QUESTION: Why does a canvas bag of water cool when the bag is slung over the bumper of a car being driven across a hot desert? [Water seeps through the canvas. The faster-moving molecules vaporize, leaving less energy per molecule behind.]

CHECK QUESTION: Name at least two ways to cool a hot cup of coffee. [You can increase evaporation by blowing on it or pouring it into the saucer to increase the evaporating area. You can cool it by conduction by pouring it into the cooler saucer or by putting silverware in it, which absorbs heat and provides a radiating antenna.]

Condensation: Evaporation is a cooling process. The opposite of evaporation is condensation, which is a warming process. After explaining this, ask why many people begin drying in the shower stall before getting out. (In the enclosed shower, appreciable warming by condensation offsets the cooling by evaporation.)

Make the point that a change of state from liquid to gas is not entirely condensation or evaporation. Both can occur together. We speak about the net effect. Make clear just what is cooling when evaporation occurs and what is warming when condensation occurs. To say that one thing cools is to say that another warms. When a hot cup of coffee cools by evaporation, the surrounding air is warmed. Conservation of energy reigns!

CHECK QUESTION: When alcohol is applied to your skin, why do you feel a chilly sensation? [Your skin is chilled by the rapid evaporation of the alcohol.] Why do you feel extra warm on a muggy day? [You are warmed by the condensation of vapor on your skin.]

If you haven't shown the collapsing can demonstration in your atmospheric pressure lecture, now is a good time to do it.

DEMONSTRATION [23-1]: Heat some aluminum soda pop cans on a burner. The can should be empty except for a small amount of water that is brought to a boil to make steam. Use a pot holder or tongs to pick up a can and quickly invert it into a basin of cold water. Crunch! The atmospheric pressure immediately crushes the can with a resounding WHOP! It is evident that condensation of the steam and vapor on the inside takes place, pressure is correspondingly reduced, and the atmospheric pressure on the outside crunches the can. Repeat the procedure, but this time invert cans into a basin of very hot water, just below boiling temperature. The can will crunch again, but less forcefully than before. Steam molecules stick to the water surface, hot or cool, like flies sticking to fly paper (Figure 23-5). Repeat the procedure again, but this time invert cans into *boiling* water. No crunch occurs. Guide your class into the explanation that the *net* effect is no change, since the amount of condensation of steam is balanced by an equal amount of vaporization from the boiling water.

Figure 23-5: A humorous way to present the condensation of water vapor is as follows. Ask: Why does a glass containing an iced drink become wet on the outside, and why is a ring of moisture is left on the table? I inject a bit of humor here and state that the reason is... and then I write a big *23-5* on the board. Ask: Why would the walls of the classroom become wet if the temperature of the room were suddenly reduced? State: The answer is... then underline your *23-5* written on the board. Ask: Why does dew form on the morning grass? State: The answer is... and put another line beneath *23-5*. Ask: Why does fog form, and how do the clouds form? Now, go back to your *23-5*. By now your class is wondering about the significance of *23-5*. Announce that you are discussing *Figure 23-5*. With class attention and interest focused, go on to discuss the formation of fog and clouds.

Boiling: Discuss boiling and the roles of applying heat and/or pressure in the boiling process. A tactic I use throughout my teaching is to ask students to pretend that they are having a conversation with a friend about the ideas of physics. Suppose a friend is skeptical about the idea of boiling being a cooling process. I tell my class that to convince the skeptic, first point out the distinction between heating and boiling. If the friend knows that the temperature of boiling water remains at 100°C regardless of the heat applied, point out that this is so because the water is cooling by boiling as fast as it is being warmed by heating. If the friend is still not convinced, ask her to hold her hands above a pot of

boiling water — in the steam. She knows she'll be burned, but burned by what? (By the steam) Where did the steam get its energy? It comes from the boiling water, so energy is leaving the water. That's what is meant by cooling!

Discuss the role of pressure on boiling and illustrate this with the pressure cooker. Explain how a geyser is like a pressure cooker. Here's an interesting tidbit. The average depth of the ocean is about two miles, and the boiling point at this depth of seawater is about 370°C (700°F). You may also wish to discuss the operation of a coffee percolator.

CHECK QUESTIONS: In the high mountains, is the time required to bring water to a boil longer or shorter than it is at sea level? [Shorter] Is the time required for cooking food longer or shorter? [Longer]

DEMONSTRATION [23-2]: Evacuate air from a flask of water that is at room temperature. Remove enough air so that the water in the flask will boil from the heat of the students' hands as it is passed around the classroom. (Use only a thick-walled flask that won't implode.)

DEMONSTRATION [23-3]: Do the triple-point demonstration in text Figure 23-10.

Freezing: Recall the open structure of ice crystals discussed in Chapter 21. This model explains why introducing foreign molecules that do not fit into the structure of ice lower the freezing point. It also explains why pressure causes regelation.

DEMONSTRATION [23-4]: If you decide to do this demonstration on the regelation of an ice cube with a copper wire, (text Figure 32-11), keep in mind that the wire must be a good heat conductor for it to work.

Energy and Changes of State: Ask if it is possible to add heat to a substance without raising its temperature. Ask why a steam burn is more damaging than a burn from boiling water at the same temperature. In answering these questions, discuss the change of state graph in text Figure 23-13 and tie this to Figure 23-12. After citing examples of a change of state where energy is absorbed, give examples where energy is released. For example, people sometimes say that it is too cold to snow. Explain that when it is snowing, the temperature of the air is higher than normal. It is actually never too cold to snow, but whenever it is snowing the air is relatively warm.

Ask: What happens in a room when the refrigerator door is left open? Tell the student to compare it to placing an air conditioner in the middle of a room instead of mounting it in a window. What would be the result if an air conditioner was mounted backwards in a window?

Further explanation may be needed for the check question on page 341 (the answer is on page 341). Explain by drawing molecules on the board as shown below, each molecule has the same KE. Ask: What happens if they bounce from each other at the same speed? (They will each have the same KE afterwards. KE before and after is the same.) Now ask : What happens if they collide and one gains speed? Show this with a larger size KE. This is okay if the other molecule has a correspondingly small KE, which you write with smaller letters. (Again, KE before and after is the same.) Now consider what happens if the molecule that loses KE is a water molecule. (If it is hit by a fast-moving molecule of any kind, it will be brought up to high speed and high KE again.) What happens if it encounters another slow-moving water molecule, one that has similarly given its energy to another molecule in collision? (The two probably stick together.) If this happens for many water molecules in a sample of gas, then the KE per molecule of remaining gas should increase as water condenses. Voilà!

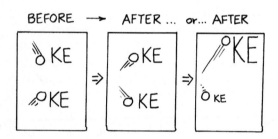

An interesting example of energy absorption during a change of state is the heat shield on spacecraft for re-entry into the atmosphere. The KE of a spacecraft in orbit is many times greater than the amount of energy needed to vaporize the craft. The shield is made of a synthetic resin or plastic ablative material that dissipates heat by melting and vaporizing. At altitudes between about 40 km and 25 km, almost the entire KE is dissipated within a period of about one minute, heating the shield to several thousand degrees Celsius. Because of its very low conductivity, only a small percentage of the heat evolved at re-entry is absorbed by the craft. A centimeter or two of the ablative material is consumed by ablation, radiating about 80% of the heat to the surrounding air. Interesting extension: If the re-entry trajectory is too steep, heating will be too severe to deal with by ablative cooling. If the trajectory is too flat, the spacecraft will be in danger of being "bounced off" the earth's atmosphere and will overshoot forever into space. Spacecraft usually enter the atmosphere at angles between about five and ten degrees to the earth's surface.

Another interesting application of change of state that is not in the text is the fine spray of water that firefighters use in combating fires. Rather than douse the burning materials with water, which may be in short supply, a fine spray that vaporizes easily is often effective in lowering the temperature to below the ignition point of the materials. This is accomplished by the energy absorbed by the tiny drops (more surface area per volume) in changing state.

Fire walking: Ask students to describe the effect on wet feet of walking over red hot wood coals. [In addition to the effect of low conductivity of wood, (whether it is cool or hot) even less energy is transferred to the feet when they are wet. Internal energy of the hot coals is used in changing the moisture to vapor. It is important that the coals be of wood, which has low conductivity. Severe burning would surely be the result of attempting to walk on red hot stones.] Caution your students to not try this, for it is very dangerous. Many people have been badly burned when conditions weren't right.

More Think-and-Explain Questions

1. Would evaporation be a cooling process if all the molecules in a liquid had the same speed?

   Answer: No, the energy of molecules leaving the liquid would be no different than the energy of molecules left behind. Although the internal energy of the liquid would decrease with evaporation, the energy per molecule would be unaffected. No change in temperature would take place.

2. Imagine that you are given a closed flask of water (one that has been checked for strength so it doesn't implode). When you hold it in your bare hands, the water begins to boil. You're impressed. Now explain why this happens.

   Answer: The air in the flask is at a very low pressure, so the heat from your hand causes the water to boil at this reduced pressure.

3. Suppose an inventor proposes a design for cookware that will allow water to boil at temperatures lower than 100°C, thus cooking food with less energy. Comment on this idea.

   Answer: It is a very poor idea, because it is the high temperature that cooks food, not the bubbles that may or may not be present in boiling water. When food cooks in boiling water, it does so because the water has a temperature of 100°C, not because it is boiling.

4. Why will spraying fruit trees with water before a frost help to protect the fruit from freezing?

   Answer: Every gram of water that undergoes freezing releases 80 calories of energy, much of it to the fruit. So while freezing occurs, the fruit is being warmed to the freezing temperature of water (which is not as low as the freezing temperature of the fruit). In a way, the cold weather freezes the water instead of the fruit. The thin coating of ice then acts as an insulating blanket against further coldness.

5. What is the relationship of text Figure 23-5 to the moisture that forms on the inside of closed windows in a stationary car on a cool night?

   Answer: The warm air generated in the car's interior meets the cold glass. The resulting decrease in molecular speed results in condensation of water on the inside of the windows.

6. Why do clouds often form above mountain peaks? (Hint: Consider the updrafts.)

   Answer: Air swept upward expands in regions of less atmospheric pressure. This expansion causes cooling, which means molecules are moving at speeds that are low enough to result in the molecules sticking together when they collide, so the moisture forms a cloud.

7. Air conditioning units contain no water. Yet it is common to see water dripping from them when they're running on a hot day. Explain.

   Answer: Condensation of moisture in the surrounding air occurs on the cool surface of these devices.

8. Why will clouds tend to form above either a flat or a mountainous island in the middle of the ocean? (Hint: Compare the specific heat capacity of the land with that of the water and the subsequent convection currents in the air.)

   Answer: Because land has a lower specific heat capacity than water does, the land is warmed faster than the surrounding water is. This causes updrafts above the warmed land; the rising air laden with H_2O expands, cools, and condenses (text Figure 24-7).

9. Why does a dog pant on a hot day?

   Answer: Dogs have no sweat glands (except between the toes for most dogs), so they must cool by the evaporation of moisture from the mouth and the respiratory track. Dogs literally cool from the inside out when they pant.

Computational Problems

1. How many calories are given off by one gram of steam at 100°C that changes state to one gram of ice at 0°C.

   Answer: 540 + 100 + 80 = 720 calories.

2. How does the heat given off by one gram of steam that condenses to boiling water compare to the heat given off by one gram of boiling water that cools to form ice, and then continues giving off energy until it reaches absolute zero. (The specific heat capacity of ice is 0.5 cal/gram°C.)

   Answer: 540 calories for one gram of 100°C steam to 100°C boiling water: For the same boiling water to absolute zero; 100 cal to cool to 0°C water, 80 cal to change to ice; 273 x 0.5 = 136.5 cal from 0°C to -273°C. In total, 100 + 80 + 136.5 = 316.5 cal, this is appreciably less than the heat given off for simply a change of state from steam to boiling water!

24 Thermodynamics

To use this planning guide work from left to right and top to bottom.

Chapter 24 Planning Guide
• The bulleted items are key: Be sure to do them!

Topic	Exploration	Concept Development	Application
Absolute Zero		• Text 24.1/Lecture Con Dev Pract Pg 24-1	Exp 63 (1 period)
First Law		• Text 24.2/Lecture	
Adiabatic Processes		• Text 24.3/Lecture Demo 24-1	
Second Law		• Text 24.4, 24.5/Lecture	Nx-Time Q 24-1
Entropy		Text 24.6, 24.7/Lecture	

Video: *None*
Evaluation: Chapter 24 Test

See the Chapter Notes for alternative ways to use all these resources.

Objectives

After studying Chapter 24, students will be able to:
- Describe the concept of absolute zero.
- State the first law of thermodynamics and relate it to energy conservation.
- Describe adiabatic processes and give examples.
- State the second law of thermodynamics and relate it to heat engines.
- Define the ideal efficiency of a heat engine in terms of input and output temperatures.
- Define entropy and give examples.

Possible Misconceptions to Correct

- There is no lower or upper limit of temperature.
- The vast internal energy of bodies like the ocean can be converted to useful energy.
- A friction-free heat engine would be a 100% efficient engine.

Demonstration Equipment

- [24-1] Pressure cooker, heat source, and pot-holder type glove

Introduction

In keeping with the preceding chapters on heat, this chapter focuses on the environment. Particular emphasis is given to the atmosphere. What do most people talk about in casual conversations? They discuss the weather, of course. This chapter presents some physics concepts involved in the weather.

The topics absolute zero and internal energy were introduced in Chapter 21 and are treated in more detail in this chapter. This chapter concludes Unit III and is not prerequisite to the chapters that follow. It may be skipped if only a brief treatment of heat is required.

Suggested Lecture

Absolute Zero and the Kelvin Scale: Follow your discussion of the Celsius and Fahrenheit scales with the concept of a lowest temperature—absolute zero and the Kelvin (K) scale.

A treatment of absolute zero is in the experiment *Uncommon Cold*. Simply state that the Kelvin scale is "nature's scale" and begins at the coldest possible value for its zero point. Note the degree symbol (°) is not used with K to reinforce the concept that its zero point was not chosen by human convention.

CHECK QUESTION: Suppose you order a piece of hot apple pie in a restaurant. The waitress brings cold pie, at 0°C, straight from the fridge. You tell her you'd like the pie to be twice as hot. When it comes back what will the temperature of the pie be? (After encouraging neighbor discussion, state that "zero degrees" is a wrong answer.) Do not give the answer yet. Ask what the new temperature would be if the pie were initially 10°C and acknowledge that the answer is not 20°C! Now you're ready for the "Celsius, the Village Tailor" story which follows.

Celsius, the Village Tailor:
Hold a meter stick against the wall of the lecture room, so that the bottom of the vertically-oriented stick is about one meter above the floor. State that you are Celsius, the village tailor, and that you measure the heights of your customers against the stick, which is fastened to the wall. State that there is no need for the stick to extend to the floor or to the ceiling, since your shortest and tallest customers fall within the extremities of the stick. Mention that all tailors using the same method would agree about the heights of their customers, providing their measuring sticks were fastened the same distance above the "absolute zero" of height. It happens that the distance to the floor, the "absolute zero," is 273 notches (the notches are the same size on the stick itself). Tell them that a very short woman enters the shop. The top of her head meets the zero mark on the measuring stick. As you take her zero reading, she comments that she has a brother who is twice her height. Ask the class to determine the height of her brother. Then ask for the temperature of the twice-as-hot apple pie. When this is understood, ask why the pie will not *really* be 273°C and why the 10°C pie will not really be 293°C. (Considerable heat has gone into changing the state of the water in the pie, which is why it is "dried out.") If you wish to avoid the change of state factor, begin your discussion with the temperature of something such as a piece of metal that will not change state for the temperature range in question.

You may wish to use the "Hotel Elevator" in which Kelvin Hall is at the basement level, Celsius Hall is on the first floor, and Fahrenheit Hall is on the third floor. Occupants of each hall measure zero height from their own respective floors.

First Law of Thermodynamics:
Introduce the first law of thermodynamics by discussing the findings of Count Rumford. When cannon barrels were being drilled and became very hot, Rumford discovered that it was the friction of the drills that produced the heating. Recall the definition of work, *force x distance*, and state that the metal is heated by the frictional force times the distance that the various parts of the drill bit move. Have your students rub their hands together and feel them warm up.

Continue by discussing Joule measuring the mechanical equivalent of heat with his paddle wheel apparatus (text Figure 24-2). Ask what happens to the temperature of a penny when you slam it with a hammer. Similar heat changes occur in water. Emphasize that the first law is simply the law of energy conservation for thermal systems.

Adiabatic Processes:
Cite the opposite processes of compression and expansion of air and how each affects the temperature of the air. It's easy to see that compressing air into a tire warms the air, and that when the same air expands through the nozzle in escaping, it cools. Have your students blow on their hands, as text Figure 24-5 suggests.

DEMONSTRATION [24-1]: Bring water to a boil in a regular pressure cooker. Then remove the weighted cap so that steam expands violently from the nozzle. For drama, put your gloved hand in the path of the "steam" about 20 cm above the nozzle. Ask if you dare do the same with a bare hand. Then remove the glove and hold your hand in the stream. Amazing! Actually the "steam" is quite cool. Explain that your hand is not in the steam, which is invisible in the first one-to-three centimeters above the nozzle. Your hand is in condensed vapor, which is cooled considerably by expansion (and mixing with the surrounding air).

Discuss cloud formation that occurs as moist air rises, expands, and cools.

If you have a model of an internal combustion engine, such as the one shown in text Figure 24-4, strongly consider showing it in class and explaining its working processes.

Meteorology and the First Law:
Discuss the adiabatic expansion of air rising in our atmosphere. Ask if it would be a good idea to go for a balloon ride on a hot day wearing a T-shirt. Would it be a better idea to bring warm clothing? A glance at text Figure 24-6 will be instructive.

Discuss the check question on text page 350 that discusses yanking down a giant dry-cleaner's garment bag from a high altitude and the resulting temperature changes. Interesting stuff.

Note that there is more to Chinook winds than is discussed in the text. As text Figure 24-7 suggests, the warm moist air that rises over a mountain cools as it expands, and then undergoes precipitation as it gains latent heat energy when vapor changes state to become liquid (rain) or solid (snow). When the energetic, dry air is compressed as it descends on the other side of the mountain, it is appreciably warmer than if precipitation hadn't occurred. Without the heat given off into the air by precipitation, it would cool a certain amount during adiabatic expansion and warm the same amount during adiabatic compression, with no net increase in temperature.

Discuss temperature inversion and the role it plays in air pollution. On the matter of pollution, acid rain is creating problems in many parts of the world. Ask students to discuss the effects of acid rain on the environment as well as on human-made objects, such as automobiles.

Second Law: Begin the second law of thermodynamics by discussing Think and Explain Question 7 on text page 360, where a hot tea cup is immersed in cold water. Stress that if the cup were to become warmer at the expense of the cold water becoming cooler, the first law would not be violated. You're on your way with the second law.

According to a U. S. Patent Office worker, the greatest shortcoming of would-be inventors is their lack of understanding of the laws of thermodynamics. The Patent Office has long been besieged with schemes that promise to circumvent these laws. This point is worth discussing. Point out Carnot's efficiency equation and its consequences, for example, why better fuel economy is achieved when driving on cold days.

CHECK QUESTION: Temperatures must be expressed in kelvins when using the formula for ideal efficiency, but it may be expressed in either Celsius or kelvins for Newton's law of cooling. Why? [In Carnot's equation, ratios are used; in Newton's law of cooling, only differences.]

CHECK QUESTION: Incandescent lamps are typically rated as 5% efficient, and fluorescent lamps are rated only 20% efficient. Now we say they are 100% efficient. Isn't this contradictory? [They are 5% and 20% efficient as light sources, but 100% efficient as heat sources. All the energy input, including light energy, becomes heat very quickly.]

Entropy: Conclude your treatment of this chapter with your best ideas on entropy — the measure of messiness.

More Think-and-Explain Questions

1. On a cold 10°C day, your friend who likes cold weather says she wishes it were twice as cold. Taking this to mean she wishes the air had half of its internal energy, what temperature would this be?

   Answer: The air will be at half its absolute temperature, or (1/2)(273 +10) = 141.5K. To find how many Celsius degrees below 0°C this is, we first subtract 141.5K from 273K. This is 273 – 141.5 = 131.5K below the freezing point of ice, or -131.5°C. (Or simply, 141.5 – 273 = –131.5°C.) Quite nippy!

2. Is it possible to construct a heat engine that produces no thermal pollution? Explain.

   Answer: According to the second law of thermodynamics, it is not possible to construct a heat engine that is without exhaust. If the exhausted heat is undesirable, then the engine is a polluter. If the exhausted heat is desirable, for heating a swimming pool for example, the heat engine produces no thermal pollution.

3. What happens to the efficiency of a heat engine when the temperature is lowered in the exhaust reservoir?

   Answer: When the temperature of the reservoir is lowered, efficiency increases; substitution of a smaller value of T_{hot} into $(T_{hot} – T_{cold})/T_{hot}$ will confirm this. (Re-express the equation as $1 – T_{cold}/T_{hot}$.)

Computational Problems

1. To increase the efficiency of a heat engine, would it be better to increase the temperature of the reservoir while holding the temperature of the sink constant, or to decrease the temperature of the sink and hold the resevoir's temperature constant. Show work.

   Answer: As in Question 3 above, inspection shows that decreasing T_{cold} will contribute to a greater increase in efficiency than will be contributed by increasing T_{hot} by the same amount. For example, let T_{hot} equal 600K and T_{cold} equal 300K. Then efficiency = (600K– 300K)/600K = 1/2. Now let T_{hot} be increased by 200K. Now efficiency = (800K–300K)/800K = 5/8. Compare this with T_{cold} decreased by 200K, in which case efficiency = (600K–100K)/600K = 5/6, which is clearly greater.

2. What is the ideal efficiency of an automobile engine wherein fuel is heated to 2700K and the outdoor air is 300K?

   Answer: Ideal efficiency is (2700K – 300K) /2700K = 24/27 = 88.8%.

25 Vibrations and Waves

To use this planning guide work from left to right and top to bottom.

Chapter 25 Planning Guide

• *The bulleted items are key: Be sure to do them!*

Topic	Exploration	Concept Development	Application
Vibrations	• Act 64 (1 period)	• Text 25.1/Lecture Demo 25-1	
Waves in General		• Text 25.2, 25.3, 25.4/Lecture Con Dev Pract Pg 25-1	
Transverse Waves	Act 66 (1 period)	• Text 25.5/Lecture	
Longitudinal Waves		• Text 25.6/Lecture	
Interference	Act 67 (>1 period)	• Text 25.7/Lecture	Exp 65 (>1 period)
Standing Waves		• Text 25.8/Lecture	Nx-Time Q 25-1
Doppler Effect		• Text 25.9/Lecture	
Bow and Shock Waves		• Text 25.10, 25.11/Lecture Con Dev Pract Pg 25-2	Nx-Time Q 25-2

Video: *Vibrations and Waves*
Evaluation: Chapter 25 Test

See the Chapter Notes for alternative ways to use all these resources.

Objectives

After studying Chapter 25, students will be able to:
• Relate a drawing of a sine curve to the crest, trough, amplitude, and length of a wave.
• Describe the relation between the frequency and the period of a wave.
• Describe what it is that travels when a wave moves outward from a vibrating source.
• Describe what affects the speed of a wave.
• Distinguish between a transverse wave and a longitudinal wave.
• Distinguish between constructive and destructive interference.
• Define a standing wave and explain how it occurs.
• Describe the Doppler effect for sound and relate it to the blue and red shifts for light.
• Describe the conditions necessary for a bow wave to occur.
• Describe the conditions necessary for a sonic boom to be heard.

Possible Misconceptions to Correct

• Wave speed and wave frequency are synonymous.
• When a wave travels in a medium, the medium moves with the wave.
• Wave amplitude and wave displacement are synonymous.

- Combinations of waves can be added but not cancelled.
- Changes in wave speed, rather than changes in wave frequency, constitute the Doppler effect.
- A sonic boom is a momentary burst of high pressure produced when something exceeds the speed of sound, rather than being a continuous front of high pressure generated by faster-than-sound sources.

Demonstration Equipment

- [25-1] Simple pendulum, and meter stick
- [25-2] Thin ruler or meter stick
- [25-3] Loose-coil spring and/or rope to shake.
- [25-4] Telephone torsion-type wave machine (the type distributed about 20 years ago by Bell Telephone Company)

Introduction

This chapter provides necessary background for the following chapter, as well as useful background for all the chapters in Unit IV.

As the chapter stands, it contains a multitude of terms and different ideas. In order to avoid information overload, note that torsional waves, although important, are not discussed.

A note of caution: When toying with a pendulum, it is easy to forget the distinction between a simple pendulum and a physical pendulum. A simple pendulum is one where the mass of the bob is very small compared to the length of string. Its rotational inertia is simply ml^2, where m is the mass and l the length of the pendulum. However, if the mass is not concentrated at the end of a string but makes up part of a stick, we no longer have a simple pendulum. Such is the case with a physical pendulum, where rotational inertia is less. The rotational inertia of a meter stick swung from one end, for example, is $1/3\ ml^2$ (text Figure 11-12), and pivoted about its midpoint, it's $1/12\ ml^2$. Many good teachers forget this distinction when making lab calculations of pendulum-bob speed at the bottom of its swing. For a simple pendulum, $v = \sqrt{2gh}$, but for a physical pendulum, the speed is greater. A physical pendulum is less lazy for its mass than a simple pendulum is. Recall that rotational inertia is greatest when all the mass involved is at the maximum distance from the rotational axis.

This idea is nicely illustrated by comparing each fall of a pair of upright meter sticks with their lower ends against a book on your lecture table — one meter stick is bare and the other has a ball of clay or some other mass attached to its upper end. Because of the greater rotational inertia per mass for the clayed stick, it is slower to rotate to the table. And while we're on falling sticks, compare the falling time of a meter stick that rotates with its lower end against a book to a stick that is allowed to slide across the table as it falls. Careful investigation will show that the sliding stick reaches the table first (its center of mass falls nearly vertically), while the stick with its lower end against the book must rotate and travel a longer distance. This is similar to time comparisons for a mass sliding down an incline and a mass dropping the same vertical distance.

Two computer simulations from the *Laserpoint* disk *Good Stuff!* are recommended. The simulation *Longitudinal Waves* should be shown when you introduce the same in lecture. Waves travel across the screen while the medium moves back and forth in place. The simulation *Doppler Effect* shows the decreasing pitch of sound when a plane or car passes an observer. When it moves faster than sound, a shock wave like the bow wave of a boat is produced.

Suggested Lecture

Vibrations: Begin by tapping your lecture table. Call attention to how frequently you tap and relate this to the term *frequency*. Tell students to note time intervals between taps and to relate this to the term *period*. Establish the reciprocal relationship between frequency and period before you begin vibrations and waves.

> DEMONSTRATION [25-1]: Attach a small massive weight to a piece of string about one meter long and swing it to and fro. This is a simple pendulum. Identify frequency, then period. Time how long it takes to swing to and fro ten times. Repeat to show the result does not change from trial to trial. (Galileo is credited as being the first to report this.) Divide the time by ten to get the period (or skip division and use the total time for the comparisons).

Now ask the class if changing the weight at the end of the pendulum will change the period. After they have given various answers, add more mass to the end of the string without changing the overall length of the pendulum. Repeat the swing process and show the same result as before. Weight does not affect the period of the pendulum. (Weight does not affect the rate of free fall or the rate of sliding down a frictionless surface. Likewise, as text Figure 25-1 suggests, pendulums of the same length will have the same period.)

> CHECK QUESTION: What principle of mechanics accounts for the different periods of pendulums of different length? [Rotational inertia, see Chapter 11]

DEMONSTRATION [25-2]: Show an upside-down pendulum by holding up a long thin ruler or meter stick and swinging it to and fro. Show that its period depends on pendulum length.

Interestingly enough, buildings behave the same way. Each building has its own period. Each of the two towers of the World Trade Center in New York City has a period of ten seconds. On a windy day the towers sway to and fro in 10-second cycles, swinging as much as one meter on a side in a strong wind. If the gusts come in rhythm with the vibrations of the building, resonance occurs. (This idea is discussed again in the next chapter.) Shorter buildings have shorter periods. The period of a 20-story building may be 1.5 seconds.

Wave Description: Vertically move a piece of chalk up and down on the board, tracing and retracing a straight vertical line. Call attention to how "frequently" you oscillate the chalk. Relate this to the definition of frequency. Also discuss the idea of displacement and amplitude (maximum displacement). With appropriate motions, show different frequencies and different amplitudes. Then do the same while walking across the front of the board tracing out a sine wave on the board. Show waves of different wavelengths.

DEMONSTRATION [25-3]: You and a student each hold the end of a stretched spring and send transverse pulses along it. Stress the idea that only the disturbance, rather than the medium, moves along the spring. Shake the spring to produce a sine wave. Then send a stretch-and-squeeze motion (elongation and compression) down the spring to show a longitudinal pulse. Send a sequence of pulses, and you have a wave. After some discussion, produce standing waves in the spring.

CHECK QUESTION: With respect to the direction of the wave's motion, how do the directions of vibrations differ for transverse and longitudinal waves? [Sideways, or perpendicular, for transverse waves; along the spring, or parallel, for longitudinal waves]

DEMONSTRATION [25-4]: Show waves on a torsion-type wave machine. These once were distributed by the Bell Telephone Company.

Cite the sameness of the frequency of a vibrating source and the frequency of the wave it produces. Explain, or derive, that wave velocity equals frequency times wavelength. Support this with examples, first with the freight car question on text page 367 and then with the water waves as in text Think and Explain Questions 2, 3, and 4. Calculate the wavelength of a local popular radio stations. (For example, for 1000 kHz on the dial, wavelength = speed/frequency = $(3 \times 10^8 \text{m/s})/10^6 \text{Hz}$ = 300 meters. That's surprisingly long!) If you discuss electromagnetic waves, be sure to contrast them with longitudinal sound waves and distinguish between the two.

Electromagnetic Waves (Optional): Depending on the design of your course, consider discussing Chapter 27 and Chapter 37 material and exploring the family of electromagnetic waves and the way they group according to wavelength and frequency. (Refer ahead to text Figure 27-5.) You can use local radio stations as examples and discuss such things as assigned frequency, clear channel, wattage, and directional signals. You can bring in well-known waves such as microwaves, CB's, black light, X rays, and gamma rays. Or, you can wait until later to discuss such waves.

Interference: Describe interference by drawing text Figure 25-10 on the board. If you have a ripple tank, show the overlapping of water waves and interference. Produce standing waves.

CHECK QUESTION: Can waves overlap in such a way as to produce a zero amplitude? [Yes, that is the destructive interference characteristic of all waves.]

Make a pair of transparencies of concentric circles. Superimpose them on your overhead projector and show the variety of interference patterns that result when their centers are displaced. One example is in text Figure 25-12.

You'll return to interference of sound in the Chapter 26, Section 10 when you demonstrate that bringing a pair of out-of-phase radio speakers face to face results in sound cancellation. Or, you could do this demonstration now.

Doppler Effect: Introduce the Doppler effect by throwing a sponge rubber or plastic foam ball around the room. Before doing so, place in the ball an electronic whistle that emits a sound of about 3000 Hz. Or, you can swing a sound source in a horizontal circle at the end of a string. Relate the sound that results to the siren of a fire engine and radar of the highway patrol (see text Figures 25-17 and 25-18). Note that sound requires a medium; radar doesn't.

Refer to text Figure 25-15 to explain the Doppler effect. Draw circles to show the top view of circular ripples made by a bug bobbing in the water. Wave speed is the same in all directions, as is evident by the circular shape. Wave frequency is the same in all directions also, since wavelength and speed are the same in all directions. Now consider a moving bug and the pattern it makes (text Figure 25-16). Explain how the frequency of waves is increased in front of the bug. Waves would be encountered more often (more frequently) if your hand was in the water in front of the bug. Similarly waves would be encountered less often (less frequently) in back of the bug. Likewise with the waves from the moving sources of text Figures 25-17 and 25-18.

> CHECK QUESTION: The waves are more crowded in front of the swimming bug and more dragged out behind. Is the wave *speed* greater in front of the bug and less behind the bug? [No, no, no! Frequency, not speed, is greater in front of the bug and less in back. Emphasize the distinction between wave speed and wave frequency.]

Bow Waves and Shock Waves: Ask the class to consider the waves made by two stones thrown in the water. Sketch overlapping waves as shown in the sketch below. Ask where the water is highest above the water level, and then use Xs to indicate the two places where the waves overlap. This is constructive interference. From this illustration and your swimming bug, you can extend the bug to speeds greater than wave speeds and show the regions of overlap that produce the bow wave (sketch Figures 25-15, 25-16, and 25-19). Next show how a bow wave makes up the V-shaped envelope with a series of overlaps. Go on to discuss the shock waves produced by supersonic aircraft.

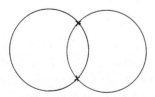

The analogy between bow waves in water and shock waves in air is useful. Questions raised by students about shock waves and the sonic boom can be effectively answered by translating the question from one of an aircraft in the air to one of a speedboat knifing through the water; a situation much easier to visualize. For example, if you're enjoying a picnic lunch at the edge of a river and a speedboat comes by and douses you, you're not apt to attribute the dousing to the idea that the speedboat just exceeded the speed of the water waves. You know the boat is generating a continuous bow wave as long as it travels faster than waves do in water. The same is true for flying aircraft.

CHECK QUESTION: Why is it that a subsonic aircraft, no matter how loud it may be, cannot produce a shock wave or sonic boom? [There will be no overlapping of spherical waves to form a cone unless the craft moves faster than the waves it generates.]

The treatment of shock waves is simplified in the text. There are actually two parts of a shock wave, the outer cone which is the superposition of condensations and an inner low-pressure cone which is the superposition of rarefactions. A graph of pressure vs. time forms an "N", as shown in the sketch below. Sonic boom damage is intensified with the incidence of low pressure rapidly following the high-pressure front.

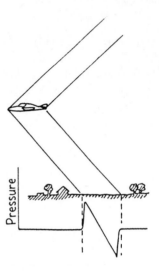

You may wish to explain shock waves in greater detail than the text does. If so, consider explaining how the speed of an aircraft can be estimated by the angle of its shock wave. Explain that shock waves are visible, since light is refracted in passing through the denser air. If possible, show any of the popular photographs of shock waves made by speeding bullets. Construct a shock wave on the board using the following sequence. Place your chalk on the board anywhere to signify time zero. Draw a meter-long horizontal line at the right to represent how far an aircraft moved in a certain time. Suppose it moves at twice the speed of sound, or at Mach 2.

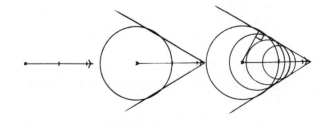

Conceptual Physics Teaching Guide

During the time the aircraft moves the distance of your one meter, the sound it made initially moved half this distance. Mark this on the midpoint of your line. State that the initial sound has expanded spherically, which you represent two-dimensionally by drawing a circle as shown on the sketch. Explain that this circle represents only one of the nearly infinite circles that make up the shock wave, which you draw by making tangents from the end point to the circle. The shock wave should be a 60 degree wedge; 30 degrees above your horizontal line and 30 degrees below it. Now move the center ten centimeters at a time in the direction of travel and draw circles (reduce the radius each time) within the two tangents. Explain how the speed of the craft is simply the ratio of the horizontal line (1 m) to the radial distance (1/2 m) of the big circle, and likewise for the respective horizontal lines to radii of smaller circles. (If your students are science students, at this point and not before, introduce the sine function). Now you can test of all this. Distribute Practice Pages from the *Concept Development Practice Book*, Chapter 25, and have your students construct shock waves of different angles to find the speeds of craft by investigating the generated angles. Or, you can construct a shock wave of different angles on the board and ask your class to estimate the speed of the craft that generated it. (This is featured on Next-Time Question 25-2.) In making constructions, the most common student error is constructing the right angle from the horizontal line, rather than from the shock wave line that is tangent to the circle. Have your students draw circles for other Mach numbers. Be sure to practice all this several times. Making these geometrical constructions is an enjoyable activity!

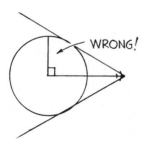

More Think-and-Explain Questions

1. What is the source of wave motion?

   Answer: Something that vibrates

2. If you triple the frequency of a vibrating object, what happens to its period?

   Answer: The period decreases to 1/3 its former value.

3. How far, in terms of wavelength, does a wave move in one period?

   Answer: One wavelength

4. What kind of motion would you apply to a spring to produce a transverse wave? A longitudinal wave?

Answers: Apply to-and-fro motion at right angles (90° and 270°) to the spring's axis; to and fro (0° and 180°) along the axis of the spring.

5. A stone is tossed into a quiet pond and waves spread over its surface. What becomes of the energy when they die out?

   Answer: The energy is transformed into thermal motion of the water molecules (internal energy).

6. Whenever you watch a high-flying aircraft overhead, its sound seems to come from behind the craft, rather than from where you see it. Why?

   Answer: It takes a noticeable amount of time for the sound to travel between the aircraft and you. The sound you hear was generated when the aircraft was farther back.

Computational Problems

1. What is the frequency, in hertz, that corresponds to each of the following periods: (a) 0.10 s, (b) 5 s, (c) 1/60 s, (d) 24 hours?

   Answers: (a) $f = 1/T = 1/0.10$ s $= 10$ Hz; (b) $f = 1/5$s $= 0.2$ Hz; (c) $f = 1/(1/60)$ s $= 60$ Hz; (d) $f = 1/24$ h x 1 h/3600 s $= 1.16$ x 10^{-5} Hz.

2. What is the period, in seconds, that corresponds to each of the following frequencies: (a) 10 Hz, (b) 0.2 Hz, (c) 60 Hz, (d) 1.15 x 10^{-5} Hz?

   Answers: (a) 0.10 s, (b) 5 s, (c) 1/60 s, (d) 86 400 s (24 hours)

3. The crests on a long surface water wave are 20 m apart, and in 1 minute 10 crests pass by. What is the speed of this wave?

   Answer: $v = f\lambda = 10$/min x 20 m $= 200$ m/min, or 3.3 m/s.

4. If the speed of a longitudinal wave is 340 m/s and the frequency is 1000 Hz, what is the wavelength of the wave?

   Answer: Wave length $= (340$ m/s)/1000 Hz $= 0.34$ m.

5. At one place in a groove of a phonograph record the "wave" is 0.01 cm long. This travels past the stylus with a speed of 40 cm/s. (a) What is the frequency of vibration in the stylus? (b) In an inner groove where the speed is 20 cm/s, what will be the wavelength to give the same frequency as in (a)?

   Answers: (a) 4000 Hz; (b) 0.005 cm; (a) $f = v/\lambda = (40$ cm/s)/ 0.01 cm $= 4000$ Hz. (b) $\lambda = v/f = (20$ cm/s)/4000 Hz $= 0.005$ cm; half as much.

To use this planning guide work from left to right and top to bottom.

Chapter 26 Planning Guide

• The bulleted items are key: Be sure to do them!

Topic	Exploration	Concept Development	Application
Origin and Nature of Sound	Act 68 (<1 period)	• Text 26.1, 26.2, 26.3/Lecture Demos 26-1, 26-2, 26-3, 26-4, 26-5	
Speed of Sound		• Text 26.4/Lecture Demo 26-6	Nx-Time Q 26-1
Forced Vibrations		• Text 26.5/Lecture	
Natural Frequency		• Text 26.6/Lecture	
Resonance		• Text 26.7/Lecture Demo 26-7	Exp 69 (1 period)
Interference		• Text 26.8/Lecture Demos 26-8 – 26-10 Con Dev Pract Pg 26-1	
Beats		• Text 26.9/Lecture Demo 26-11	

Video: *Sound*
Evaluation: Chapter 26 Test

See the Chapter Notes for alternative ways to use all these resources.

Objectives

After studying Chapter 26, students will be able to:
- Relate the pitch of a sound to its frequency.
- Describe what happens to air when sound moves through it.
- Compare the transmission of sound through air with its transmission through solids, liquids, and a vacuum.
- Describe factors that affect the speed of sound.
- Give examples of forced vibration.
- Describe the conditions for resonance.
- Describe the conditions for beats.

Possible Misconceptions to Correct

- The speed of sound is the same in all media.
- Sound cannot cancel sound.
- Resonance and forced vibrations are the same thing.

Demonstration Equipment

- [26-1] Large tuning fork and container of water
- [26-2] Large bare loudspeaker and audio oscillator (or other source to drive speaker)
- [26-3] Aluminum rod, about one meter long
- [26-4] Rosin or pitch to put on fingers for stroking aluminum rod (sources: gym rosin or Canada balsam, which is refined pine pitch used in biology labs)
- [26-5] Vacuum jar with ringing doorbell inside and vacuum pump
- [26-6] Pair of matched tuning forks mounted on sounding boxes

- [26-7] Small bare loudspeaker and music source, and a baffle board (about one square meter of cardboard) with a hole somewhat smaller than the speaker cut in middle
- [26-8] Speaker box to hold the bare speaker
- [26-9] Stereo tape player with mono mode, matching speakers, and switch or jacks to reverse the polarity of one of the speakers
- [26-10] Laser and a pair of mounted tuning forks, see Demonstration 26-7

Introduction

This chapter lends itself to many interesting lecture demonstrations: a ringing doorbell inside a vacuum jar that is being evacuated; the easily-seen vibrations of a tuning fork illuminated with a strobe lamp; the interference when a pair of radio speakers are out of phase with each other; resonance and beats from a pair of tuning forks mounted on sound boxes; and the 8-mm film loop, *Tacoma Narrows Bridge Collapse*.

Forced vibrations, resonance, and interference for sound provide a useful background for the chapter on light. The concepts introduced for sound are applied later to light.

An impressive demonstration of vibrations and sound is the stroking of an aluminum rod with your fingers. It's necessary to put some pine pitch or violin-bow rosin on the rod and/or on your fingers. (If you use pitch, acetone dilutes it so you can get a light layer on the rod and your thumb and finger.) Hold the rod by one end for the fundamental tone, then at the midpoint for the first harmonic. Practice is highly recommended.

Another neat demonstration is resonance in a long stovepipe. Put three layers of wire window screen on crossed wires one quarter of the way up from the bottom of the stovepipe. Heat over a Bunsen burner, but take care not to melt the screen! The screen is a white-noise generator, while the tube is the frequency selector that resonates at the fundamental frequency of the tube. When removed from the flame, the pipe apparatus continues to sound as the wire screen cools. Turn the pipe sideways for a few seconds until the sound subsides, then turn it back to a vertical position — the sound returns. Very impressive!

Be sure to play music via stereo speakers in your class. Play monophonic music from a tape cassette recorder or from a radio. First play one speaker, then play both in phase. The resulting sound is only slightly louder. This illustrates the logarithmic nature of our response to changes in sound level, since doubling the intensity of the sound results in only a small perceived increase in loudness. It takes about eight times the intensity to perceive twice the loudness.

Speaker systems for stereos have polarity indicators on their terminals, which is essential for optimum sound. The speakers must operate with the same relative phase for a given input signal; that is, both diaphragms must move in the same direction for a given direction of current from the amplifier. If they move in opposite directions, which happens when they are out of phase, destructive interference results. The sound is clearly lower in volume when the speakers are out of phase. Go one spectacular step further and face the speakers against each other. Near silence! Then pull the speakers farther apart and the sound returns. This demonstration is a must.

The diminishing of sound by destructive interference may suggest a cancellation of energy. Not so, if the input is a vibration and the output is induced voltage, the radio loudspeaker is in effect a microphone. This is electromagnetic induction, and is analogous to a motor as a generator and vice versa, depending on input and output. When the speakers face each other they "drive" each other, inducing back voltages in each other that cut the currents down in each. Thus, energy is diminished, but it is not cancelled.

Interestingly enough, electronic synthesizers are so sophisticated now that human voices are often produced electronically in recording studios, thus replacing backup singers. The voices of some recording artists are so electronically altered that they are not the same quality as they are in live performances. The recorded performance is a hard act for the artist to follow. Your students will probably have more information about this in class discussion.

There are three programs on the computer disk *Good Stuff!* that nicely complement the chapter. *Sum of Two Waves* illustrates interference, wave superposition, and beats. *Frequency Analyzer* treats the different sounds of instruments. To see the overtones that give sounds their character, you will need an analog-to-digital converter, a microphone, and an audio amplifier. Sing or play a steady note and adjust the pitch to obtain the clearest display. The computer program, *Music*, turns the keyboard of the Apple into a piano keyboard. Although you can only play one note at a time, you will see the note you're playing on the music staff displayed on the screen.

Suggested Lecture

Sound: Begin on a light note and state with mocked profundity that sound is the only thing the ear can hear. Then state more seriously that the source of sound or any wave motion is a vibrating object.

> DEMONSTRATION [26-1]:   Tap a large tuning fork and show that it is vibrating by dipping the vibrating prongs into a cup of water. The splashing water shows that the prongs are moving. (Small tuning forks do not work well.)

> DEMONSTRATION [26-2]:   Show a large radio speaker without its cover. Play low frequencies with an audio oscillator (or other source) so that students can gather around and actually see the diaphragm vibrating. This is most impressive the first time it is seen!

> DEMONSTRATION [26-3]:   Hold an aluminum rod (one meter or so in length) horizontally at the midpoint and strike one end with a hammer. You will create vibrations that travel and reflect back and forth along the length of the rod. The sustained sound heard is the due to energy "leaking" from the ends, about 1% with each reflection. So at any time the sound inside is about 100 times as intense as that heard at the ends. (This is similar to the behavior of light waves in a laser.) Shake the rod to and fro in order to illustrate the Doppler effect.

> DEMONSTRATION [26-4]:   Rub some pine pitch or rosin on your fingers and stroke the aluminum rod. If you do it properly, it will "sing" very loudly. Do this while holding the rod at the midpoint and then at different places to demonstrate harmonics. (Of course you practiced doing this first!)

> DEMONSTRATION [26-5]:   Show the ringing doorbell suspended in a bell jar that is being evacuated of air (see text Figure 26-7).

Media that Transmit Sound: While the loudness of the ringing doorbell sound diminishes, discuss the movement of sound through different media, such as gases, liquids, and solids (use the vibrating Ping-Pong ball analogy). Ask why sound moves faster in warm air then it does in cool air. [Faster-moving balls take less time to bump into one another.]

Speed of Sound: Discuss the speed of sound through different media. Sound moves four times as fast in water as it does in air, and it moves about eleven times as fast in steel. The elasticity of these materials, rather than their densities, accounts for the different speeds. Cite the fact that the Native Americans used to place their ears to the ground in order to listen for distant hoof beats.

Sound travels faster in moist air than it does in dry air. Why? [Because H_2O molecules move faster than N_2 or O_2 molecules do. This shortens the time between the collisions that transmit sound energy.] Ask why H_2O molecules move faster. Use the "check your neighbor" routine to see if you can prod your students into applying some good physics here. [H_2O molecules have less mass (18 amu) than O_2 (32 amu) and N_2 (28 amu) molecules do. At the same temperature, all the air molecules have the same kinetic energy, so the less massive ones move faster. That is because less m means more v for the same KE in $KE = 1/2\ mv^2$. Or by the principle of exaggeration: If an elephant and a mouse run with the same KE in order to do equal work when colliding with a barn door, which of the two must be running with the greater speed?]

Discuss the speed of sound and how to estimate one's distance from a thunder storm. ($d = vt = 340$ m/s $\times t$)

Forced Vibration, Natural Frequency, and Resonance: Introduce the phenomenon of forced vibration by striking an unmounted tuning fork and then holding its base firmly against your lecture table or the chalkboard. Next perform the same procedure using the mechanical part of a music box. Turn the handle when it is held in air and then place it on your table or chalkboard. This is impressive.

Acknowledge the natural frequencies of the prongs of the music box "comb," and the prongs of different tuning forks, as well as other objects around the room. Compare the sounds of a couple of pennies dropped on a hard surface — one dated before and one after 1965. The older penny is made of 95% copper and 5% zinc and sounds noticeably different than does the newer pure zinc core pennies that are clad with copper. The ear can discriminate between more than 300 000 tones!

> DEMONSTRATION [26-6]:   Show resonance with a pair of tuning forks. Explain how each set of compressions from the first fork pushes the prongs of the second fork in rhythm with its natural motion. Compare this to pushing somebody on a playground swing. Illuminate the forks with a strobe light for the best effect!

When you are adjusting the frequency of one of your tuning fork boxes by moving the weights up or down the prongs, call attention to the similarity between this adjustment and tuning a radio receiver. When one turns the radio knob to select a station, one is adjusting the frequency of the radio set to resonate with the frequency of incoming station signals.

Cite other examples of resonance: the chattering vibration of a glass shelf when a radio placed on it plays a certain note; the loose front end of a car that vibrates only when the car goes certain speeds; a crystal wine glass that shatters when a singer hits a certain note; troops that march in step when crossing a bridge. Conclude your treatment of resonance with the exciting film loop *The Tacoma Narrows Bridge Collapse.*

(This is a good place to break.)

Interference: Review interference by sketching overlapping sine curves on the board (as in text Figure 26-13, or more simply like Figure 25-10 in the previous chapter). Now you're ready for a series of fantastic demonstrations — perhaps the most unforgettable of your course!

> DEMONSTRATION [26-7]: Play music via a very small naked speaker, a few centimeters in diameter, connected to the auxiliary output of a portable cassette recorder. The music will sound tinny. Then produce a baffle (large flat piece of cardboard) with a hole slightly smaller than the size of the speaker cut in its middle. Place the speaker behind the hole and note how the sound quality is much improved. The baffle reduces the interference between the back and front waves.

Explain that a radio loudspeaker produces waves from both the front and the back that are 180° out of phase. When it is producing a compression in front, it is producing a rarefaction in back, and vice versa. When sound reaches your ears from both the front and back of a speaker, destructive interference occurs. This is most pronounced for long waves where the different distances traveled from the speaker to you are relatively small. (Sound will both diffract around the speaker and reflect from surfaces behind the speaker. Diffraction is enhanced for long waves.) The result is that a naked loudspeaker tends to sound tinny because it produces little sound energy for wavelengths that are much longer than its diameter. The long-wavelength base notes are cancelled. This cancellation is notably reduced when the baffle is introduced.

> DEMONSTRATION [26-8]: Now place the same naked speaker behind the hole in a small closed box to show even better sound quality. The speaker enclosure is a so-called "infinite baffle," which prevents rear waves from interfering with front waves.

Most popular speaker enclosures are more complex than simple boxes are. Whatever the features of the enclosures are, your class now knows why speakers are mounted in them!

> DEMONSTRATION [26-9]: Play monophonic music from a cassette recorder (or any monophonic amplifier) via a pair of common enclosed stereo speakers. Place the speakers side by side facing the class. (1) Play one speaker. (2) Play both in phase. The resulting sound is slightly louder than with one speaker. (3) With a switch or otherwise, reverse the leads to one of the speakers to reverse the phase. The sound is much less intense than it is from the single speaker. (4) With speakers still facing forward, try different separation distances and illustrate the wavelength dependence of the interference. You'll increase the loudness of the sound heard by increasing the distance between the speakers. Sound with wavelengths greater than the distance between speakers is cancelled. Waves shorter than the separation distance are not totally cancelled. Note the variations in the quality of the sound heard. (5) Now for the grand finale. Face the speakers toward each other with only a small gap between them. Play one speaker and then both in phase. No big deal. Now reverse the polarity of one of the speakers. The result is almost total silence. Sound at virtually all wavelengths is being cancelled by destructive interference. As you move them apart and increase the distance between them, shorter wavelengths avoid total destructive interference and the sound level increases. Spectacular!

Beats: Interference and beats are shown nicely with an oscilloscope trace of sound from a pair of sound sources slightly out of sync.

> DEMONSTRATION [26-10]: Show beats as Paul Robinson does, by bouncing laser light off a pair of vibrating tuning forks. Quite lively!

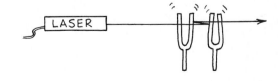

More Think-and-Explain Questions

1. Why do flying bees buzz?

   Answer: Bees buzz when in flight because they flap their wings at audio frequencies.

2. Why is an echo weaker than the original sound?

   Answer: An echo is weaker than the original sound, because sound spreads with distance and is therefore less intense. If you are at the source of the echo, the echo will sound as if it originated on the other side of the wall from which it is reflected (just as your image in a mirror appears to come from behind the glass). The sound is weaker still because the wall does not reflect sound perfectly.

Computational Problems

1. What is the wavelength of a 680-Hz tone in air? What is the wavelength of a 68 000-Hz ultrasound wave in air?

   Answer: Wavelength = speed/frequency = (340 m/s)/680 Hz = 1/2 m: For a 68 000-hertz wave; wavelength = (340m/s)/68 000 Hz = 0.005 m = 1/2 cm.

2. A bat flying in a cave emits a sound pulse and receives its echo in one second. How far away is the cave wall when the bat emits the signal?

   Answer: 170 m: Assuming the speed of sound to be 340 m/s, the cave is 170 meters away. This is because the sound took 1/2 second to reach the wall (and 1/2 second to return). Distance = speed × time = 340 m/s x 1/2 s = 170 m.

3. A sea vessel surveys the ocean bottom with ultrasonic sound that travels at 1530 m/s in seawater. How deep is the water if the time delay of the echo from the ocean floor is 4 seconds?

   Answer: The ocean floor is 3060 meters deep. The 4-second time delay means that the sound reached the bottom in 2 seconds. Distance = speed × time = 1530 m/s × 2 s = 3060 m.

4. A rule for estimating the distance in kilometers between an observer and a lightning strike is to divide the number of seconds in the interval between the flash and the thunder by 3. Is this rule correct?

   Answer: The rule is correct. The speed of sound in air (340 m/s) can be rounded off to 1/3 km/s. Then, from distance = speed x time = (1/3 km/s) × (number of seconds), note that the time in seconds divided by 3 yields the same value.

27 Light

To use this planning guide work from left to right and top to bottom.

Chapter 27 Planning Guide

• *The bulleted items are key: Be sure to do them!*

Topic	Exploration	Concept Development	Application
Early Concepts		Text 27.1/Lecture	
Speed of Light		• Text 27.2/Lecture Con Dev Pract Pg 27-1	
Electromagnetic Waves		• Text 27.3/Lecture	Nx-Time Q 27-1
Transparent Materials		• Text 27.4/Lecture	
Opaque Materials		• Text 27.5/Lecture	
Shadows	Act 70 (1 period)	• Text 27.6/Lecture	Exp 71 (1 period)
Polarization	Act 72 (1 period)	• Text 27.7/Lecture Con Dev Pract Pg 27-2 Demos 27-1, 27-2, 27-3	Nx-Time Q 27-2

Video: *Light and Color*
Evaluation: Chapter 27 Test

See the Chapter Notes for alternative ways to use all these resources

Objectives

After studying Chapter 27, students will be able to:

- Describe the dual nature of light.
- Explain why it is much more difficult to measure the speed of light than the speed of sound.
- Describe the relation between light, radio waves, microwaves, and X rays.
- Explain what happens to light when it enters a substance and how the light frequency affects what happens.
- Describe the conditions for solar and lunar eclipses.
- Give evidence to show that light waves are transverse.
- Explain why polarized sunglasses are helpful in cutting the glare of the sun from horizontal surfaces such as water and roads.

Possible Misconceptions to Correct

- Light and sound have the same wave nature but they have different frequencies.
- Light is fundamentally different than radio waves, microwaves, and X rays.
- Light passes through transparent materials in a way similar to bullets passing through materials.
- Polarized light is composed of a unique kind of light waves.

Demonstration Equipment

- [27-1] Rubber tubing and a grate (from an oven or refrigerator)
- [27-2] Three polarized lenses (about 30 cm by 30 cm each)
- [27-3] Slide projector and slides made of polarized material, crumpled cellophane, and/or shiny cellophane tape (both of which discern polarized light)

Introduction

Some teachers begin their physics course with light, a topic that has a greater appeal for many students than mechanics does. Your course could begin with this chapter and continue through the following chapters of Unit IV, or you could pick up Chapters 25 and 26 before finishing the rest of this unit. Be flexible. The reason for starting with Chapter 27 is to avoid intimidating students with the more technical nature of Chapter 25. This sequence, Chapters 27, 25, 26, 28, 29, 30, 31, is a gradual entrance to the study of physics. If Chapter 27 is used as a launch point, only the definitions of speed and frequency need to be introduced. In addition, give a demonstration of resonance with a pair of tuning forks. Your students will then experience resonance, which will be a jumping off place to understanding the interaction of light and matter.

Note the "depth of the plow" in the treatment of light transmission, reflection, and absorption. The aim is not to separate and name these categories, but to get into the physics. Your students will get into some rather deep physics in this chapter. It is a depth they can understand. Understanding more than one expects and discovering more than one thought was there is a real joy of learning. So this should be an enjoyable chapter — the reason some teachers may opt to begin their course here.

In reference to the visual illusions in text Figure 27-23 on text page 408, the vertical lines are parallel and the tiles are not crooked. This can be seen by looking at the page at an angle. The width of the hat is the same as its height, the "fork" and "rectangular" piece could not be made in the shop, and there are two THEs in the sign reading PARIS IN THE THE SPRING.

The equally bright rectangles illustrate a phenomenon called *lateral inhibition*, wherein gradual differences in the intensity of light aren't perceived very well. The human eye can perceive differences in brightness that range from about 500 million to one. The difference in brightness between the sun and the moon, for example, is about 1 million to one. Lateral inhibition prevents the brightest places in the visual field from outshining the rest. Whenever a receptor cell on your retina sends a brightness signal to your brain, it also signals neighboring cells to dim their responses. In this way, you even out your visual field and you can discern detail in both bright and dark areas. Lateral inhibition works to exaggerate the difference in brightness at the *edges* of places in your visual field. Edges, by definition, separate one thing from another. At an edge, differences rather than similarities are accentuated. The gray rectangle on the left in text Figure 27-23 appears dimmer than the one on the right when the edge that separates them is in view. But cover the edge between them with a pencil or your finger, and they look equally bright. That's because both rectangles *are* equally bright. Each rectangle is shaded lighter to darker, moving from left to right. Your eye concentrates on the boundary where the dark edge of the left rectangle joins the light edge of the right rectangle, and your eye-brain system assumes that the entire left rectangle is darker.

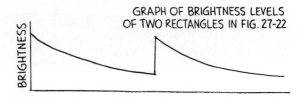

GRAPH OF BRIGHTNESS LEVELS OF TWO RECTANGLES IN FIG. 27-22

Suggested Lecture

Begin by stating that, strictly speaking, light is the only thing we see. To understand what light is, we will first try to understand how it behaves. Call attention to the rainbow of colors that are dispersed by a prism or by raindrops in the sunlight. We know white light can be spread into a spectrum of colors. Ask your students to consider the world viewed by little creatures who could only see a tiny portion of the visible spectrum; creatures who would be color-blind to all the other visible frequencies of light. Their world view would be very limited. Then state that we are like those little creatures in that the spectrum of colors we can see is a tiny portion of the electromagnetic spectrum — less than one tenth of one percent! We are color-blind to the other parts of the electromagnetic spectrum. The instruments of science have extended our vision, however. These instruments are not microscopes and telescopes that enable closer viewing of the part of the spectrum familiar to us. It is the infrared detecting devices, microwave and radio receivers, that allow us to explore the lower-frequency end of the spectrum, and it is ultraviolet, X-ray, and gamma-ray detectors that let us "see" the higher-frequency end. What we see without unaided eyes is a tiny part of what's out there in the world around us.

Electromagnetic Spectrum: Depart from the order in the book and call student's attention to text Figure 27-5, the electromagnetic spectrum, and the tiny part that is visible light.

> CHECK QUESTION: Where does sound fit in the electromagnetic spectrum? [It doesn't! Sound is a mechanical wave, a vibration of material particles. It too has a spectrum, only a small part of which is heard! Electromagnetic waves are not vibrations of material stuff; they are vibrations of pure energy. (We will see in Chapter 37 that the energy is in the form of electric and magnetic fields — hence the name *electromagnetic waves*.)]

Speed of Light: Light is fast — too fast to measure over short distances with everyday devices. However, over long distances, it can be measured without sophisticated equipment. Discuss Roemer's calculation (text page 396).

CHECK QUESTION: What would be the consequences if Michelson had used a four-sided rotating mirror rather than the octagonal mirror to measure the speed of light? [It would have had to spin twice as fast because it would need to rotate 1/4 turn rather than 1/8 turn to direct the returning light into the eyepiece.]

Light and Transparent Materials: Ask students to recall the earlier demonstration of sound resonance. (If you haven't done this, demonstrate resonance with a pair of tuning forks how.) Sound resonance then becomes the basis for understanding the interaction of light with matter. In some cases light strikes a material and rebounds, a phenomenon called *reflection* (Chapter 29). In other cases, where light continues through the material, we call the material *transparent.*

Point out the value of scientific models in understanding physical phenomena. Hence the text discussion of tuning forks and imaginary springs that hold electrons to the nuclei of atoms. A model is not correct or incorrect, but rather it is useful or nonuseful. Models must be refined or abandoned if they fail to account for various aspects of a phenomenon.

Go over the explanations on text page 401, and Figure 27-8.

CHECK QUESTION: Why is light slower in transparent materials such as water or glass? [According to the model treated in the text, there is a time delay between the absorption of light and its re-emission. This time delay serves to decrease the average speed of light in a transparent material. Similarly, the average speed of a basketball moving down a court depends on the holding time of each player.]

Opaque Materials: State that light generally has three fates when it is incident upon a material (1) the light bounces off (reflects), (2) the light is transmitted through the material, and (3) the light is absorbed by the material. Usually a combination of all three fates occurs. When absorption occurs, the vibrations given to electrons by incident light are often great enough to last for a relatively long time, during which the vibratory energy is shared by collisions with neighboring atoms. The absorbed energy warms the material.

CHECK QUESTION: Why is a black tar road hotter to the touch than a pane of window glass in the sunlight is? [Sunlight is absorbed and turned into internal energy in the road surface, but it is transmitted through the glass to somewhere else.]

Shadows: Illustrate the different shadows cast by small and large sources of light. Ask students why no definite shadow appears when they hold their hands above their desks. Relate this to the multiple sources and diffused light in the room. Discuss solar and lunar eclipses after the lab activity *Shady Business*.

CHECK QUESTION: Does the earth cast a shadow in space whenever a lunar or solar eclipse occurs? [Yes, but not only when these events occur—the earth, like all objects illuminated by light from a concentrated source, casts a shadow. Evidence of this perpetual shadow is seen at these special times.]

Polarization: Discuss how to distinguish between polarized and nonpolarized light.

DEMONSTRATION [27-1]: Tie a rubber tube to a distant firm support and pass it through a grate (from a refrigerator or oven shelf). Have a student hold the grating while you shake the free end and produce transverse waves. Show that when the grating axis and the plane of *polarization* are aligned, the wave passes through the grate. When they are at right angles to each other, the wave is blocked.

DEMONSTRATION [27-2]: Cross a pair of polarized lenses in front of a light source, as shown in the left and center photos in text Figure 27-20. Show this schematically with vectors on the chalkboard.

The explanation for sandwiching a third polarized lens, as shown in the right photo of text Figure 27-20, is not given in the text. The question is again raised in Appendix D, where the student is asked to come to you as a last resort — after his or her own efforts. The explanation is shown here with vectors (and in the answer to Next-Time Question 27-2).

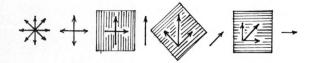

For an ideal polarizer, 50% of the incident light (electric) vectors are transmitted. That is why a sheet of polarized material passes so little light compared to a sheet of window pane. The transmitted vectors are aligned and the light is polarized. So in the above diagram, 50% of the incident light vectors are transmitted by the first sheet, 50% cos θ gets through the second sheet, where φ is the angle between the polarization axes of both sheets; and (50% cos θ) cos φ gets through the third sheet, where θ is the angle between the polarization axes of the second and third sheet. The emerging vector represents the amplitude of the emerging light. For a value of 45° for both angles, the amplitude that emerges is 25% of the incident amplitude. Since the intensity of any wave is proportional to the square of the amplitude, the intensity emerging would be 0.25^2 or 6.25% of the incident intensity. This, combined with the fact that polarizers are less than ideal, is why an image is barely seen through the three-sheet system.

An important goal of this course is to help students make distinctions between things.

Reflected Polarized Light: Explain how the light that reflects from nonmetallic surfaces is polarized parallel to the plane of the surface. Do this by using the analogy of skipping flat rocks off the surface of the water, when the plane of the rock is parallel to the water surface. Then draw on the chalkboard the sets of sunglasses shown in the boxed question on text page 407. Ask which sets are the best for reducing road glare. If you want to discuss viewing three-dimensional slides and movies, you'll have a transition for doing so by using sunglasses C.

The following demonstration can be presented now as a follow up to polarization, or it can be saved until you treat interference in Chapter 31. Either way, it can be one of the most memorable demonstrations of your course.

DEMONSTRATION [27-3]:   The vivid colors that emerge from cellophane between crossed polarized lenses make a spectacular demonstration. Have students make up some slides of cut and crinkled cellophane mounted on polarized material, which can be obtained inexpensively from a science materials supplier. Place the slides in a slide projector and rotate a sheet of polarized material in front of the projecting lens so that a changing montage of colors is displayed on the screen.

Also show color slides of the interference colors seen in the everyday environment, as well as those colors in microscopic crystals. This is more effective with two projectors with hand dissolving from image to image on the screen. Do this in rhythm to some music and you'll have an unforgettable lecture demonstration!

This demonstration neatly ushers you into the next chapter, the study of color!

More Think-and-Explain Questions

1. If a one-sided, silvered plane mirror were used in the Michelson apparatus, how much faster would it have to spin so that reflected light would be seen in the telescope?

   Answer: Only twice as fast! When light incident at 45 degrees on it makes a round trip, the mirror must rotate 90 degrees in order to be at 45 degrees to the telescope. Unless the back side of the mirror also reflects, the next pulse of light would not occur until the back side of the mirror turns through 270 degrees.

2. Pretend a person can walk only at a certain pace — no faster, no slower. If you time her uninterrupted walk across a room of known length, you can calculate her walking speed. If, however, she stops momentarily along the way to greet others in the room, the extra time spent in her brief interactions gives an *average* speed across the room that is less than her walking speed. How is this like light passing through glass? In what way is it not like it?

   Answer: The person walking across the room and pausing to greet others is analogous to the transmission-of-light model in that there is a pause with each interaction. However, the same person who begins the walk ends the walk. This is where the analogy breaks down. In light transmission, there is a "death-birth" sequence of events as light is absorbed and "new light" is emitted in its place. The light to first strike the glass is not the same light that finally emerges. (Another analogy is a relay race, or mail going to San Francisco from Kansas City by Pony Express — a change of wands or horses occurs at each station.)

3. A lunar eclipse is always that of a full moon. That is, the full moon is always seen just before and after the earth's shadow passes over it. Would it be possible to have a lunar eclipse when the moon is in its crescent or half-moon phase? Explain.

   Answer: No, a full moon occurs every 28 days when the sun, earth, and moon are lined up. When the alignment is exact, an eclipse occurs. A solar eclipse occurs when the moon is between the sun and earth. A lunar eclipse occurs when the earth is between the sun and moon, and the moon is located in the earth's shadow. When the moon is in a half-moon stage, it is one quarter of a month away from being in the earth's shadow, and it is even farther from the shadow when it is in its crescent stage.

4. What astronomical event would observers on the moon see at the time the earth was undergoing a lunar eclipse?

Answer: These observers would see the earth in the path of the sunlight and they would witness a solar eclipse.

5. What astronomical event would observers on the moon see at the time the earth was undergoing a solar eclipse?

Answer: Moon observers would see a small shadow of the moon slowly move across the full earth. The shadow would consist of a dark spot (the umbra) surrounded by a not-as-dark circle (the penumbra).

6. Why do polarized sunglasses reduce glare, whereas nonpolarized sunglasses simply cut down on the total amount of light reaching our eyes?

Answer: Glare is composed largely of light polarized in the plane of the reflecting surface. Most glaring surfaces are horizontal (roadways, water), so sunglasses with vertical polarization axes filter the glare of horizontally-polarized light. Conventional nonpolarizing sunglasses simply cut down on overall light transmission by reflection and/or absorption.

To use this planning guide work from left to right and top to bottom.

Chapter 28 Planning Guide			
• The bulleted items are key: Be sure to do them!			
Topic	**Exploration**	**Concept Development**	**Application**
The Color Spectrum		• Text 28.1/Lecture	
Color by Reflection		• Text 28.2 /Lecture	
Color by Transmission		• Text 28.3 /Lecture	
Sunlight		• Text 28.4 /Lecture	
Mixing Colored Lights	Demos 28-1, 28-2	• Text 28.5 /Lecture Con Dev Pract Pg 28-1	Nx-Time-Q 28-1
Complementary Colors		• Text 28.6 /Lecture	
Mixing Pigments		• Text 28.7 /Lecture	
Blue Sky, Red Sunsets, Green Water		• Text 28.8 – 28.10/Lecture • Demos 28-3, 28-4	
Atomic Spectra		• Text 28.11 /Lecture Demo 28-5	Exp 73 (> 1 period)
Florescence (Optional)		Demo 28-6	
Video: *Light and Color* **Evaluation:** Chapter 28 Test			

See the Chapter Notes for alternative ways to use all these resources

Objectives

After studying Chapter 28, students will be able to:

• Explain why white and black are not colors in the same sense that red and green are colors.

• Describe why the interaction of light with the atoms or molecules of a material differs for different frequencies.

• Describe the factors that determine whether a material will reflect, transmit, or absorb light of a particular color.

• Explain how color television screens are able to display pictures in full color, even though the television tube produces only spots of red, green, or blue light.

• Define complementary colors and give examples of pairs of them.

• Distinguish between color mixing by subtraction and color mixing by addition.

• Explain why the sky is blue and why it changes color when the sun is low in the sky.

• Explain why water is greenish blue.

• Explain what the lines in a line spectrum represent and how such a spectrum can be used to identify the presence of an element.

Possible Misconceptions to Correct

- White and black are colors.
- Red, yellow, and blue light make white light.
- Red and green light make brown.
- Colored light and colored pigments, when mixed in the same combination, produce the same resulting color.

Demonstration Equipment

- [28-1] Singerman Color Apparatus (a box that casts three circles of light on a translucent screen, includes an assortment of colored filters) or its equivalent
- [28-2] Three lamps, red, green, and blue, that can be clamped to a lecture table
- [28-3] Two trays of tuning forks for simulating light sources
- [28-4] Transparent container of water and powdered milk, and a light source that gives off a strong beam
- [28-5] Either a single lecture-size diffraction grating or enough small ones to pass one out to each student, and an assortment of gas discharge tubes with an appropriate lamp to display them (atomic spectra)
- [28-6] Ultraviolet lamp and an assortment of objects that will both fluoresce and phosphoresce when they are illuminated

Introduction

The model used in the text to explain color is the *oscillator*. In the model, electrons of an atom are forced to vibrate by the oscillations of light waves. (There is a slight difference between *oscillate* and *vibrate*. *Vibrate* usually refers to mechanical motion of matter. *Oscillate* usually refers to the motion of electrons and the electromagnetic field.)

Be sure your students have seen and heard a demonstration of resonance with tuning forks. Resonance is a central idea in this chapter, and the tuning fork model accounts for selective light reflection and transmission.

If you have not already done so, show a box with a white interior that has a hole in it, as shown in Figure 22-14 on text page 323.

This is a "must-do" demonstration. Mount three floodlights on your lecture table (red, green, and blue) of shades such that when all three overlap, they produce white on a white screen. Colored bulbs of 75 watts work fine. Reduce the room lighting and stand in front of the lamps, which are illuminated one at a time to show the interesting colors of the shadows. Impressive!

The chapter ends with a very brief treatment of atomic spectra. It is placed at the end as an extention of color. Be sure to do the lab experiment *Flaming Out* with it. If you wish to continue further and cover fluorescence and phosphorescence, an optional lecture follows. Atomic spectra leads quite nicely into Unit VI, *Atomic and Nuclear Physics*. Jumping from this chapter directly to Chapter 37 has some merit if a short course is desired.

This interesting chapter can be taught very rigorously, or it can be a plateau where physics is fun. I recommend the latter. Except for atomic spectra, it is not a prerequisite to the chapters that follow.

Suggested Lecture

Display different colored objects while you discuss the oscillator model of the atom and the ideas of forced vibration and resonance as they relate to color. Discuss the color spectrum and the nonspectral "colors" white and black. Point out that the colors of nonluminous objects are the colors of the light they reflect or transmit. Discuss text Figures 28-6 and 28-7.

Sunlight: The source of all light is accelerating electrons. The internal energy of the sun shakes its electrons so violently that waves of energy bathe the solar system and extend beyond it. This is sunlight, which has a wide frequency range. Draw the radiation curve for sunlight on the board (see text Figure 28-8). Go further than the text and divide the visible portion into thirds; a low-frequency section that averages to red, a middle-frequency section that averages to green, and a higher-frequency section that averages to blue. These three regions correspond to the three regions of color to which our retina is sensitive. Hence we see that red, green, and blue are the three primary colors of white light.

The eye is most sensitive to yellow green, which is the peak frequency of sunlight. That is why today's fire engines are yellow green in color. Our eyes are also most sensitive to the yellow-green light of sodium lamps. This means that, for a given wattage, we see more under sodium light than we see under the white light of an incandescant source.

Mixing Colored Lights: The primary colors indicated in text Figure 28-9 are shown in color in Plate 1.

DEMONSTRATION [28-1]: Use the Singerman Color Apparatus (or its equivalent) to show the overlapping of the primary colors. Next show the complementary colors.

CHECK QUESTIONS: Ask several questions such as those at the top of text page 419 that refer to your display of Figure 28-9.

DEMONSTRATION [28-2]:   With the room darkened, stand between a white surface and three lamps (red, green, and blue) that are placed at least a meter apart. Students understand this better if you turn the lamps on one at a time and discuss what is seen with each added light. Turn on the red lamp and view your black shadow. Then turn on the green lamp to show that the black shadow cast by the red light is now illuminated by green light (as if the red light were casting a green shadow — strange at first). Note that the shadow cast by the green light is not black like the room is when the lights are off, but it is red like the red that was there before the green light was switched on! Most interesting, note the yellow everywhere where red and green overlap. Next turn on the blue light; the yellow background becomes white. Now you have a third shadow. Can your students account for the colors of your shadow?

Mixing Colored Pigments: Now we address color mixing as it relates to one's early finger painting experience; blue + yellow = green, red + yellow = orange, red + blue = purple. This was probably the only information on color mixing ever given to your students. Distinguish between the colors that we see as a result of absorption (color mixing by subtraction) and the effect of superposing colored lights (color mixing by addition). Discuss text Figure 28-11. Pass a magnifying glass around and look at the cyan, magenta, and yellow dots that make up the colors on the color plates in this chapter and in various magazines. A fact not mentioned in the text is that the eye can distinguish nearly 8 million differences of color.

There are no blue pigments in the feather of a blue jay. Instead there are in the barbs of its feathers tiny alveolar cells that scatter light — mainly high-frequency light. So a blue jay is blue for the same reason the sky is blue — light is scattered. Interestingly enough, although brown eyes in people are due to pigments, the blue in blue eyes is due to the scattering of light from tiny spheres in the iris of the eye.

DEMONSTRATION 28-3:   *Blue-Sky Lecture Skit* — Here's where your showmanship comes in handy. Place a variety of six tuning forks at one end of your lecture table, which you designate as the *sun* end. Name the tuning forks "red," "orange," "yellow," "green," "blue," and "violet." Ask your class what "color" sound they would hear if you struck the tuning forks in unison. Your class should answer "white." Then

suppose you have a mirror device around the forks so that when you "light" (strike) them again, a beam of sound travels down the length of your lecture table. Ask what "color" they will hear. Several might say "white" again, but state that if there is no medium to scatter the beam, they will hear nothing (unless, of course, the beam is directed toward them). Now place a tray of tuning forks at the opposite end of your lecture table, which you designate as the *earth* end (the tray can be simply a piece of wooden 2 x 4 about one-third meter long, with a dozen holes drilled in it to hold tuning forks of various sizes). Ask your class to imagine that the ends of your lecture table are 150 million km apart (the distance between the earth and the sun). State that your tray of assorted tuning forks represents the earth's atmosphere. Point to the tuning forks while calling out their "colors;" "blue," "violet," "blue," "blue," "red," "blue," "violet," "blue," "green," "blue," "violet," and so forth emphasizing the preponderance of "blue" and "violet" forks. Your tray of forks is perpendicular to the imaginary beam from the sun and represents the "thin" atmosphere seen by the sun at midday. Walk to the *sun* end of the table and again pretend to strike the forks in order to show how the beam travels down the table to intercept and scatter from the atmospheric tuning forks in all directions. Ask what "color" the class hears, and you have a blue sky, especially if they're a bit deficient in hearing violet.

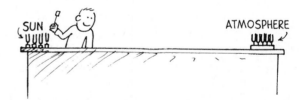

Red Sunset: Sketch text Figure 28-14 on the board to show that at sunset the sunlight must go through many kilometers of air in order to reach an observer and that blue light is scattered all along these kilometers. Ask what frequencies survive. Then go back to your sun and earth forks on the lecture table. Select a student from the class to sit in back of the tray of earth forks. Tell the class that your volunteer represents an earth observer at sunset. Go back to the sun forks and pretend to strike them. Explain that down the table comes the beam, which you follow. It hits the earth's atmosphere where most of it scatters throughout the classroom. Again, ask the class what "color" they "hear." "Blue," is the correct answer. Now you ask your volunteer what color he or she heard. "White," is the correct answer! The thin noontime atmosphere did little to the white beam from the sun. Fine! Now rotate the tray of forks representing the earth's atmosphere 90 degrees to simulate a thicker atmosphere. Go back to your sun and repeat the process. When the light reaches the earth tuning forks, ask your class what color they hear. (Although blue is still the correct answer, many will say orange, which is not seen from their vantage point.) Emphasize that blue is still the color they see, because of the preponderance of blue forks that are seen from any angle except straight on. Now ask your volunteer what color he or she hears. "Orange," is the answer! Your demonstration has been a success. By experimenting you have proved your point. Your student volunteer has heard a composite of the lower-frequency, left-over colors after the class received most all the higher-frequency blues. Consequently, the pretty colors at sunset are left-over colors!

> DEMONSTRATION [28-4]: Shine a beam of white light through a colloidal suspension of a very small quantity of instant nonfat dry milk in water, in order to demonstrate the scattering of blue and transmission of orange light. Students will see the beam in the water turn blue at the same time they see the spot of light cast on the wall turn orange.

Point out that the sky looks blue only when it is viewed against the blackness of space. The blue is really not very bright. Astronauts see no blue when looking straight downward from orbit, because the brightness of reflected earth light overwhelms the weak blue.

The Violet Sky: Compare the molecules in the atmosphere to tiny bells. When they are "struck," they "ring" with high frequencies. They "ring" the most at violet and next at blue. We're better at "hearing" blue, so we "hear" a blue sky. On the other hand, bumble bees and other creatures with good vision in violet see a violet sky. If we were better detectors of violet, our sky would be violet.

White Clouds: Larger molecules and particles, like larger bells, ring at lower frequencies. Very large ones ring in the reds. In a cloud there is a wide assortment of particles of all sizes and they ring with all colors. Ask your class if they have any idea why clouds are white! (Cumulus clouds are composed of water droplets and are white due to the multitude of constituent particle sizes. Higher-altitude cirrus clouds are composed of ice crystals. Like snow they reflect all frequencies.) Discuss the blueness of distant dark mountains and the yellowness of distant snow-covered mountains (see the boxed questions on text page 424).

Greenish-Blue Water: Water absorbs infrared rays. It also absorbs some visible light nearer the red end of the color spectrum. Take red away from white light and you are left with the complementary color — cyan. Hence, the sea looks cyan, or greenish blue. Interestingly enough, deep beneath the ocean there is no red left in white light, so things that look red in the sunlight look black under deep water. A red crab and black crab have the same appearance on the ocean floor.

Interestingly enough, tropical waters are a deeper cyan color than cooler waters are, because warmer water is purer in the sense that it has less oxygen and dissolved air in it. The greater amount of oxygen and dissolved air in cooler water not only changes its optical properties, but it supports more organisms. This makes cooler water ecologically richer in nutrients.

(This is a good place to break.)

Atomic Spectra: State that a century ago the chemical composition of the stars was thought to be forever beyond our knowledge, but today we know as much about the stars' composition as we do about the earth's. This is because the light emitted by all things is an atomic fingerprint; it reveals the atomic sources of things. These fingerprints are atomic spectral lines (see text Figure 28-11 and in color in Plate 5).

Develop the analogy between light emission and the emission of sound by tiny bells. If the bells are made to ring all at once while they are crammed together in a box, the sound will be discordant. The same is true of light emitted by atoms that are crammed together in a solid or liquid state. There is a spread of frequencies, so we get a wide radiation curve such as that from the sun, see text Figure 28-8. The light is "smudged" and appears white. (Likewise, the sounding of a wide range of sound frequencies is called *white noise*.) When bells are far apart from one another, however, the sound they emit is pure and unmuffled. So it is with atoms in the gaseous state that emit light. The light emitted by glowing atoms in the gaseous state can be separated into discrete pure colors with a spectroscope.

> DEMONSTRATION [28-5]: Show the spectra of gas discharge tubes with a large diffraction grating (an 8½ x 11 inch sheet of plastic grating is best), or pass out small gratings to the class. Follow this up with individual student-viewing of spectral lines of discharge tubes seen with a spectroscope. The spectrum of helium gas is impressive.

Give some examples of how spectroscopes are used. Spectroscopes are tools for chemical analysis. They use minute quantities of the material to be analyzed. Tiny samples of ores are sparked in carbon arcs and the light is directed through prisms or diffraction gratings to yield the precise chemical composition of the material. Note that spectroscopes are used in fields as diverse as chemistry and criminology.

Fluorescence (Optional):   You may wish to extend the textbook material and consider discussing the stimulation of light by higher-frequency, thus more energetic light — fluorescence. Show some fluorescent materials under the light of the room. Explain the role of light from overhead lamps. Light shakes the electrons in the molecules of the fluorescent material and produces not only reflection, but also emission. The term *day-glow*, which sometimes describes fluorescent paints, refers to this emission. (As the footnote on page 427 states, there is more going on here than forced vibration of electrons in the molecules. Electrons are being dislodged from their stable orbits and kicked to higher-energy orbits. We say they are *excited*. It is when they de-excite and return to lower-energy levels that they emit light. The frequency of the emitted light is directly proportional to the difference in the energy states the electron traverses in the atom.)

CHECK QUESTION: Would higher-frequency light produce more glowing? Why? [Yes, higher-frequency light is more energetic light. (This idea will be treated in detail in Chapter 38.)]

DEMONSTRATION [28-6]:   Show fluorescent materials illuminated with an ultraviolet lamp (black light). Discuss the observations when the black light is on, and then extinguish the black light so the room is totally dark. Ask what is happening. [High-energy ultraviolet light, incident upon the fluorescent materials, knocks electrons in the material up to higher energy states. When the electrons return to lower states, the light emitted has a frequency directly proportional to the difference in energy between the upper and lower states.] Illuminate some phosphorescent materials.

Discuss phosphorescence and give common examples; watch and clock faces, light switches, and night lights. In the case of phosphorescent materials, there is a time delay between excitation and de-excitation. The delay produces the familiar afterglow of these materials. Some phosphorescent watch faces were once activated by radioactive minerals. (Such watches are not recommended. They are harmful, especially to the eyes, which are very prone to radiation damage.)

"Oral Borealis":   This demonstration makes a good home project since it must be done with a mirror in the complete darkness. Buy a roll of hard

wintergreen-flavored mints. After your eyes have adapted to the dark, bite down on a candy as you look in the mirror. The mint will spark and glitter as you chew! This form of atomic excitation is called *triboluminescence*, (from the Greek word *tribein*, meaning to rub and the Latin word, *lumin*, meaning light.) This light is created by friction. When the candy crystals are crushed, nitrogen atoms give off ultraviolet light, as well as violet and blue light. The wintergreen oil (methyl salicilate) is fluorescent and converts the invisible ultraviolet light into visible blue light adding to the effect. The excitation of nitrogen in the upper atmosphere near the earth's poles is what makes the glowing aurora borealis, so, in a sense, you have a mini-aurora in your mouth.

The same blue glow can be seen if you unroll masking tape in the dark. The blue glow occurs where the tape is unsticking from itself. This does not work with all masking tapes, however.

More Think-and-Explain Questions

1. Why are black and white not listed as colors?
   Answer: They are not parts of the color spectrum. Black is the absence of light, and white is a composite of all colors.

2. On a color television screen, red, green, and blue spots of fluorescent materials are illuminated at a variety of relative intensities to produce a full spectrum of colors. What dots are activated to produce yellow? Magenta? White?
   Answer: Red and green produce yellow; red and blue produce magenta; red, blue, and green produce white.

3. If the sky on a certain planet in the solar system was normally orange, what color would its sunsets be?
   Answer: An orange sky indicates preferred light scattering of low frequencies. At sunset when the scattering path is longer, very little low-frequency light would get to an observer. The less-scattered high-frequency light would produce a bluish sunset.

4. When white light shines on red ink dried on a glass plate, the color that is transmitted is red. However, the reflected color is not red. What color is it?
   Answer: The reflected color is white minus red, or cyan.

5. What causes the beautiful colors sometimes seen in the various burning materials in a fireplace?
   Answer: Atoms of the material are heated to glowing, wherein different kinds of atoms give off their own characteristic colors.

6. What is the evidence for the claim that iron exists in the atmosphere of the sun?
   Answer: Iron spectral lines are found in the solar spectrum.

29 Reflection and Refraction

To use this planning guide work from left to right and top to bottom.

Chapter 29 Planning Guide

• *The bulleted items are key: Be sure to do them!*

Topic	Exploration	Concept Development	Application
Law of Reflection		• Text 29.1, 29.2/Lecture Con Dev Pract Pg 29-1	
Mirrors	Act 74 (>1 period) Act 75 (1 period) Act 76 (1 period) Act 77 (1 period)	• Text 29.3 /Lecture	Exp 78 (>1 period) Nx-Time Qs 29-1, 29-2
Diffuse Reflection		• Text 29.4/Lecture	
Reflection of Sound		• Text 29.5/Lecture	
Refraction		• Text 29.6/Lecture Con Dev Pract Pg 29-2 • Demos 29-1, 29-2	
Refraction of Sound		• Text 29.7/Lecture	
Refraction of Light		• Text 29.8/Lecture	Nx-Time Qs 29-3, 29-4
Atmospheric Refraction		• Text 29.9/Lecture	Nx-Time Qs 29-5, 29-6
Dispersion in a Prism		• Text 29.10/Lecture Demo 29-3	
Rainbows		• Text 29.11/Lecture	
Total Internal Reflection		• Text 29.12/Lecture	

Video: *Reflection and Refraction*
Evaluation: Chapter 29 Test

See the Chapter Notes for alternative ways to use all these resources

Objectives

After studying Chapter 29, students will be able to:
• Distinguish between what happens to light when it strikes a metal surface and when it strikes glass or water.

• Given the direction of light striking a reflective surface, predict the path of the reflected light.
• Explain why the image formed by a mirror is a virtual image.
• Describe the conditions for diffuse reflection.
• Give examples of ways to control reflected sound.

- Explain the change in direction of a water wave when it crosses a boundary between deep and shallow water.
- Give examples of refraction of sound waves and its effects.
- Give examples of refraction of light and its effects.
- Explain how a prism separates white light into colors.
- Describe the conditions for a rainbow.
- Describe the process of total internal reflection.

Possible Misconceptions to Correct

- The (average) speed of light is constant in all materials.
- The average speed of light and instantaneous speed of light in a material is the same.

- The law of reflection is restricted to plane surfaces.
- Both sound and light travel only in straight lines.
- A prism changes (rather than separates) white light into colors.
- A rainbow is a physical thing that can be approached and grasped if you're lucky.

Demonstration Equipment

- [29-1] Glass tank of water with dye added, prism, mirror, and light source (laser)
- [29-2] Root beer mug filled with root beer (or any dark liquid), and glass tank of water
- [29-3] Rainbow sticks (shown below)

Introduction

The treatment of reflection in this chapter is brief, with scant applications to convex and concave mirrors. The treatment of refraction is supported by many examples of both sound and light. The oscillator model of the atom is used to explain refractive index. If you haven't demonstrated resonance with a pair of tuning forks on sound boxes, consider doing so now.

As a brief treatment on light, this chapter can stand on its own. In this case the behavior, rather than the nature of light, is emphasized.

Assign Activities 2 and 3 on text page 451 as home projects. These ask for the minimum size mirror needed to view a person's full-length image. Seldom are these two questions answered correctly. Emphasize that the results will be surprising, and that if the students are careful they will learn something about their image in a mirror that has likely escaped their notice all their lives. They will learn that the size of their image in the mirror is independent of their distance from it). Ask them to mark the mirror where they see the top of their head and bottom of their chin, and then compare the distance between the marks with the height of their face. And then to step back and see what effect there is (if any) with increased distance from the mirror. Most students will still miss it. The fact that distance doesn't change the answer is simply not believed by some students—in spite of the evidence to the contrary. When you discuss the answer, bring into class a full-length mirror or pass a few small mirrors among your students. It's worth the extra effort.

Connecticut physics teachers Jon Wallace and Jim Harper do ripple tank demonstrations using an inexpensive clear plastic one-piece frame that

is about 3 cm deep. Put water in it and mount it on an overhead projector. Strips of screen mesh placed around the inside edges will reduce unwanted reflected waves. Use a large-diameter wooden dowel to generate waves. A gentle roll forward followed by a quick roll backward produces a nice single pulse. You control the frequency of additional pulses. If you wish to make custom-designed shapes to show diffraction, a material called Lexan© responds well to drilling or cutting.

An explanation of why a rainbow is bow-shaped is aided with this simple, easy to construct apparatus: Stick three colored dowels into a sphere of clay, plastic foam, wood, or other material that represents a raindrop. One dowel is white, one violet, and the other red, to represent incident white light and refracted red and violet light. The angles between dowels are shown in the sketch. A student volunteer crouching in front of your chalkboard shows the class how the only drops that cast light to him or her originate in drops along a bow-shaped region.

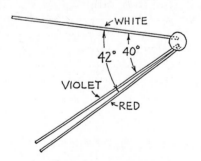

Raindrop size affects the proportions of color seen in a rainbow. Drops between 0.5 and 1 mm in diameter create the most brilliant colors. Drops of 1-2 mm diameter show very bright violet and green light with scarcely any blue. Larger drops are poor rainbow producers, because they depart from a truly spherical shape flattening by air pressure and the oscillations they undergo. In drops smaller than 0.5 mm in diameter, red is weak. In the 0.02-0.3-mm range, red is not seen. At 0.08-0.10 mm, the bow is pale with only violet vivid. Smaller drops produce weak bows with a distinct white stripe. As a sheet of polarized material will show, rainbows are polarized.

An interesting reference on rainbows is *Light and Colour*, by M. Minnaert, (Dover, 1954). An excellent set of 35-mm slides of rainbows, halos, and glories can be previewed in *Rainbows, Halos, and Glories* by Bob Greenler, (Cambridge University Press, 1980).

Lab activity, *Images*, treats the formation of virtual images. The activity, *Pepper's Ghost*, deals with multiple reflections. The experiment, *Funland*, investigates the properties of a circular spherical mirror. An application of multiple reflections is *The Kaleidoscope*. *Satellite TV* treats the focusing properties of a spherical reflector. If time is short, lab omissions for this chapter have fewer drawbacks than they do for other chapters.

Suggested Lecture

Reflection: Cite a ball bouncing on a street or against the bank of a pool table. The angle of incidence equals the angle of rebound (that is, reflection); the same rule applies to light (text Figure 29-3).

Discuss panes of glass both transmitting and reflecting light. Reflection is more noticeable when it's dark on the other side of the glass, and less noticeable when it's light on the other side. Stress that the percentage reflected (about 4%) is the same in both cases. In the first case the 4% is noticeable, but in the second case it isn't. This is like stars which are really in the sky during the day and at night.

CHECK QUESTION: How do one-way mirrors work? [They are partially silvered so they reflect about as much light as they transmit. If there is no light source behind the mirror, you cannot see the people or whatever else is there. Light reflected by the mirror overwhelms any light coming from behind it. If a lit match were behind the mirror, you could see it.]

Show by ray diagram that the image in a plane mirror is as far behind the mirror as the object is in front of it.

CHECK QUESTION: Why will a camera with a sonar or infrared range finder give poor results when you take a picture of yourself in a mirror? [The camera will focus on the mirror surface, where the sonar or infrared beam is being reflected, in order to set the automatic focus. However, your image is as far behind the mirror as you are in front of it.]

Investigate text Figure 29-6 and point out that the law of reflection along every facet of a curved surface is the same as it is for a flat surface. A single ray of light does not distinguish between a large area and the tiny part with which it interacts.

Cite the role of reflection for ghost images on broadcast TV pictures. The signal from the broadcasting station can take more than one path to reach the antenna. If it bounces off buildings and other surfaces, it may not reach the antenna in phase with the direct signal — hence the multiple image.

Here's an interesting thought: Because of the finite speed of light, your image in the mirror is always younger than you.

Diffuse Reflection: Show by a ray diagram that reflection from a rough surface is diffused. Point out that for every facet of the surface, the law of reflection holds (text Figure 29-7). We see most of our environment by diffuse reflection. Cite the different appearance of a highly waxed floor and the same floor when the wax has worn off.

Ask if students have ever noticed that when driving in a car on a rainy night, it is difficult to see the road ahead. The explanation involves diffuse reflection and is a neat one. When the road is dry, it is rough, and light from the car headlights is diffused—some light returns to the driver. But on a rainy night, the road has a water surface that acts as a mirror. The light is not diffusely reflected back to the driver, but is reflected ahead by the mirrorlike surface. This is an added disadvantage to approaching drivers, and to you. Instead of the light from headlights being diffused in many directions, it is reflected into the eyes of oncoming motorists.

CHECK QUESTION: Would your book be easier to read if the pages were shinier. Why or why not? [No, there would be more glare and less softly-diffused reflected light.]

Reflection of Sound: The law of reflection also applies to sound. Walls that diffusely reflect sound are preferred in concert halls. Point out the interesting interplay between sound and light in the disks shown in text Figure 29-12. Since both light and sound obey the same law of reflection, where it's seen is where it's heard.

Discuss echos and multiple echos—reverberations.

CHECK QUESTION: Why does your voice sound fuller when you sing in the shower? [Each note lasts slightly longer as your voice reverberates between the walls.]

Refraction of Sound: Refraction of any wave depends on changing its speed. Note the sound of a bugle being refracted both upward and downward in text Figure 29-15. The reason for the refraction shown in the figure is the change in speed through different air densities. It could also be due to different wind speeds. The upper illustration could represent faster ground winds, and the lower illustration faster upper winds. It may help to list all variables.

A useful application of sound refraction, ultrasound imaging, is replacing X rays in probing the body's internal organs. This method seems to be relatively free of dangerous side effects. Another interesting example of sound refraction is sonar.

> CHECK QUESTION: What is the key factor for refraction of any kind of wave? [A change in wave speed.]

Refraction of Light: The change in (average) speed for light was established in Chapter 27, highlighted by text Figure 27-8. Another way to explain this is as follows: If a bullet is fired through a thick board, it will emerge with a speed that is less than its incident speed. This is clear enough. It loses energy in the board, as evidenced by the warm wood around the hole and the warm bullet. State that if light did the same thing, one could direct a beam of light into a bank of glass plates, and it would slow with each successive encounter. A weak bit of light might gleam from the last plate. Ask why this doesn't happen. Recall the model of light passing through the glass plate in text Figure 27-8 (or Think-and-Explain Question 2 and its answer, on page 145 of this manual). Be sure to stress that different frequencies of light travel at different speeds in a transparent material. Red is slowest and violet is fastest (which is important information in the explanation of dispersion in a prism).

Cite the role of refraction in the atmosphere and in a prism. Note that all this applies also to sound.

> DEMONSTRATION [29-1]: Show examples of reflection, refraction, and total internal reflection with the usual apparatus; light source (laser), prisms, and a tank of water with a bit of fluorescene dye added.

> DEMONSTRATION [29-2]: Show a thick root beer mug filled with root beer or cola. Because of significant refraction of light from the glass mug to the air, the mug appears to contain more root beer than it actually does. You can show what happens by immersing an empty mug in a tank of water. Because the light doesn't change speed as much in traveling through the glass to the water, you can easily see the glass' thickness.

Total Internal Refraction: Depart from the order of the text and get into total internal refraction, as shown in your first tank-of-water demonstration. Ask your class to imagine how the sky would look from a lake bottom. [For humor, above water we must turn our heads through 180 degrees to see from horizon to horizon, but a fish need only scan twice the 48 degree critical angle to see from horizon to horizon — which is why fish have no necks!]

Fiber optics: Show some examples of "light pipes" and a lamp, as shown in text Figure 29-34. Discuss some of the many applications of these fibers, or light pipes, particularly in telephone communications. Information not in the text: Due to the higher frequencies of light compared to frequencies of electric currents, a pair of glass fibers as thin as a human hair can carry 1300 simultaneous telephone conversations, but only 24 can be carried simultaneously by a conventional copper cable. Signals in copper cables must be boosted every 4 to 6 kilometers, whereas re-amplification in lightwave systems occurs in 10-to 50-kilometer segments. For infrared optical fibers, the distance between regenerators may be hundreds or perhaps thousands of kilometers.

Call attention to the fiber-optic nature of polar bear hairs, text Figure 29-35. What's white and black and warm all under? A polar bear in the Artic sun.

The Rainbow: State that a rainbow is a complete circle, but the ground prevents our seeing the circle. Complete circles of both primary and secondary bows are sometimes visible from aircraft.

Point out that there are not distinct layers of color in the rainbow, but rather the colors merge from one hue to another. Observers on the ground separate the rainbow into the colors they have learned to identify.

> DEMONSTRATION [29-3]: Show the rainbow-sticks apparatus described earlier and compare it to the schematic drawing of a rainbow in text Figure 29-26. The white stick represents incoming white light, and the red and violet sticks represent the refracted rays. Have a student volunteer crouch in front of the board as shown in the sketch. Place the ball (the "drop") near the chalkboard so the white dowel is perpendicular to the board (from the sun at the horizon for simplicity). Position the free end of the violet dowel so that it nearly meets the volunteer's eye. State that a "drop" at this location refracts violet light to the eye. The question follows: Are there

other locations that will also refract violet light to the eye? Move the "drop" to other locations along the board while keeping the white dowel perpendicular to the board. It is easy to see that refracted violet from drops farther away miss the eye altogether. The only locations that send violet light to the eye are along a bow — which you trace with violet or blue chalk. This is easy to do if the student holds the end of the violet dowel near one eye while you scribe the arc in compass fashion.

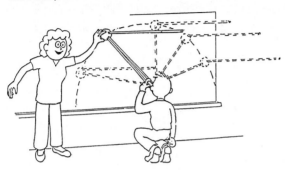

CHECK QUESTION: With the ball and dowels positioned at the top of the bow, ask where the volunteer must look to see red—above or below the violet? [Above, 2° to be exact] Show this by moving the "drop" up, whereupon the red dowel lines up with the eye. Complete your demonstration by sweeping this wider bow with red chalk.

Enlist a second volunteer to crouch in front of the board at a different position. Show how this volunteer must look to different drops in the sky in order to see a rainbow. Ask the class if the two volunteers see the same rainbow. [No, each sees his or her own personal rainbow!] Marshall Ellenstein quips that when one person says to another, "Look at the beautiful rainbow," an appropriate response is, "Move over and let me see!"

Rainbows cannot be seen when the sun is more than 42 degrees above the horizon, because the bow is below the horizon where no water drops are to be seen. Hence, rainbows are normally seen early and late in the day. We don't see rainbows at midday in summer in most parts of the world (except from an airplane, where they are seen in full circles).

Point out a significant yet commonly unnoticed feature about the rainbow—that the disk segment bounded by the bow is appreciably brighter than the rest of the sky. This is clearly shown in text Plate 6. The rainbow is similar to the chromatic aberration around a bright spot of projected white light.

Explain how the phenomenon of the halo around the moon is similar to a rainbow. The halo is produced by refraction of moonlight through ice crystals in the atmosphere. However, there are several important differences. Both refraction and internal reflection produce rainbows, whereas only refraction produces halos. To see a rainbow, the observer is between the sun and the drops. However, the ice crystals that produce halos are between the observer and the moon. Moonlight is refracted through ice crystals high in the atmosphere—evidence of the coldness up there, even on a hot summer night.

More Think-and-Explain Questions

1. What kind of wind conditions would make sound more easily heard at long distances?

   Answer: When the wind traveling toward the listener is faster above the ground than at ground level, the second waves are refracted downward, as shown in text Figure 29-16.

2. When a fish looks upward at an angle of 45 degrees does it see the sky, or does it see the reflection of the environment underneath it? Defend your answer.

   Answer: Because 45° is less than the critical angle (48°) in water, the fish sees out into the air above. The view of the upper environment seen by the fish is bounded by the edge of a large circle, which is part of a 96° cone. A photographer's "fisheye lens" is so named because it compresses a wide-angle view in a similar manner.

30 Lenses

To use this planning guide work from left to right and top to bottom.

Chapter 30 Planning Guide

• *The bulleted items are key: Be sure to do them!*

Topic	Exploration	Concept Development	Application
Features of Lenses		• Text 30.1/Lecture Demo 30.1	
Images and Rays	• Act 79 (1 period) • Act 83 (<1 period) Act 84 (1 period)	• Text 30.2, 30.3, 30.4/Lecture Con Dev Pract Pgs 30-1, 30-2	Exp 81 (>1 period) Exp 82 (1 period)
Optical Instruments		• Text 30.5/Lecture	
The Eye		• Text 30.6/Lecture Demos 30-2, 30-3	Nx-Time Q 30-1
Vision and Lens Defects	• Act 80 (<1 period)	• Text 30.7/Lecture	

Video: *None*
Evaluation: Chapter 30 Test

See the Chapter Notes for alternative ways to use all these resources

Objectives

After studying Chapter 30, students will be able to:
• Distinguish between a converging and diverging lens.
• Distinguish between a real image and a virtual image formed by a lens.
• Given the focal length of a converging or diverging lens and the position of an object, construct a ray diagram that shows the position of the image.
• Give examples of how some optical instruments use lenses.
• Explain how the human eye focuses light.
• Explain the causes of nearsightedness, farsightedness, and astigmatism.
• Give examples of aberration in lenses.

Possible Misconceptions to Correct

• A lens can produce (rather than simply concentrate) energy.
• A lens is necessary for image formation.

Classroom Equipment

• Several converging (and diverging) lenses to pass around the class

Demonstration Equipment

• [30-1] Prism and light source
• [30-2] Model of eye (see sketch)
• [30-3] Brightly-colored cards (3 or 4 different colors)

Introduction

This chapter is an extension of the previous chapter. It applies refraction to lenses and introduces the ideas of ray optics. It may be omitted without consequence to the chapters that follow.

Just as it would be futile for a swimming coach to discuss the techniques of swimming without performing the activity of swimming, it is futile to show the techniques of ray diagram construction without having students construct one. And just as swimming motions are best made in water, your students should have a lens in hand, an object, and an image to observe when they make their constructions. Ray diagrams are abstractions. Unless they are tied to experience, they are of questionable value. Let your treatment of this chapter be a class activity.

It is important that you try to do the activity on solar image with the class. The students must calculate the sun's diameter by measuring the sun's image through a pinhole. It is one of the most impressive activities in this course. It is a striking example of how the richness in life comes from not just looking at the world with wide-open eyes, but from knowing what to look for. How nice that your students can go home and announce to their parents that they calculated the sun's diameter with the use of a meterstick! Here's another example of the most profound concepts in physics being the simplest.

Begin your study of lenses with the lab activity, *Camera Obscura*, followed by *Lensless Lens*. Do the lab activity, *Air Lens*, only if you have ample time. Do the experiment *Bifocals* only after your lecture and discussion of the chapter. The lab experiment, *Where's the Point?*, makes a fun follow-up if you have a laser.

Suggested Lecture

Begin with the following demonstration.

> DEMONSTRATION [30-1]: Show how a prism refracts a ray of light. Do this with an actual prism, and then introduce ray tracing on the board.

Contrast the deviation of light by a prism with the absence of deviation by a pane of glass. Show that a pane only *displaces* light rays, and show how thicker panes produce greater displacements. Note the thick panes of glass used in aquariums. Show how a lens behaves as a smooth curved prism (text Figure 30-1).

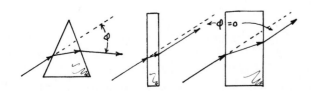

Image Formation: Confine the first part of your presentation to converging lenses. Pass some converging lenses around class and have students cast images of the windows onto the wall or the image of the ceiling lights onto their desks. (Most find this fascinating.) Have them cast images of closer brightly-illuminated objects (bright enough to cast a clearly-seen image).

> CHECK QUESTION: Is there a relationship between the image distance (distance from lens to place where sharp image appears) and the object distance (from object to lens)? [Yes, the farther the object is, the nearer the image is. Students will learn that when an object is sufficiently far away, the image appears in the focal plane.]

Point to any part of a blank wall. Ask if there is an image on the wall of some particular bright object. (Identify one on the other side of the room.) It so happens there is — over *every* part of the wall. We don't see the image, however, because it is overlapped by an infinite number of images of everything else in sight. The fascinating thing is that if the overlapping images were blocked, the image of the object in question would be seen. A light-tight barrier with a pin hole does just that. This is possible because the pin hole, by virtue of its tiny size, prevents the passage of light rays that overlap.

Have the class pass around a box with a pin hole and transluscent viewing screen on opposite ends. Describe with a simple ray diagram how images are formed and why they are inverted. Discuss the pinhole camera activity at the back of the chapter. Note that the size of the hole compared to the distance to the viewing screen affects two things; the amount of light that makes the image and the sharpness of the image. A larger hole will admit more light, but overlapping images through different parts of the hole reduce sharpness. Of course, a large hole covered by a lens solves both problems. When this was discovered and cameras utilized lenses instead of pinholes (as they first did), the greater amount of light meant pictures could be taken faster — hence the name "snap shots."

Call attention to the round spots of light that are found beneath trees on a sunny day. This is worth bringing your class outside to see — really! Just as the cultivated ear appreciates music not really heard by others, and just as the trained touch of a physician feels irregularities beneath the skin that others miss, and just as the cultivated pallet tastes food that others are insensitive to, we can see here, that the trained eye sees what others miss. The spots of light beneath the tree brances are indeed circles if the sun is overhead, or they are stretched out circles (ellipses) if the sun is low in the sky. They are literally images of the sun. This is more convincing at the time of a partial solar eclipse. During a partial solar eclipse the images of the sun are crescents. What is happening is that openings between leaves in the trees behave as pinholes when the openings are small compared to their distances from the ground. As a result, they cast images of the sun on the ground. (Class interest at this point should be in high gear.)

Poke a hole in a piece of cardboard and hold it in the sunlight. Note that in its shadow there is a circle of light on an area beneath the hole. To convince your students that the circle is not merely the image of the hole itself, cut various shapes—squares, stars, whatever. If they are held an appropriate distance away, the images remain circular. You can tell that the spot of light is truly an image of the sun when a partial solar eclipse occurs. Then the image is a crescent. At the time of a partial solar eclipse, an array of crescents rather than ellipses fall on the ground beneath trees in the sunlight!

Now for a fantastic quantitative exercise. Hold up a meter stick and announce that measurements of the elliptical images of the sun and their distances from their pinholes, together with the knowledge that the sun is 150 000 000 km from the earth, enables one to calculate the sun's diameter. To simplify this task, place a coin beneath the pinhole and adjust the distance of the pinhole from the coin so it is exactly eclipsed by the sun's image (the short diameter of the ellipse should equal the diameter of the coin). This allows an accurate method of measuring the diameter of the solar image, since the diameter of the coin is easily measured. Here's the neat part. The ratio of the image diameter to the pinhole distance is the same as the ratio of the sun's diameter to 150 000 000 kilometers. We have one equation, one unknown. [Careful measurements will show the ratio of coin size to distance from the pinhole is about 1/108. It is then easy to show that the diameter of the sun is 1/108 of 150 000 000 km. The answer of approximately 14 000 km agrees with the accepted value (shown in the text Table B-1 in Appendix B).]

CLASS PROJECT: Have your class calculate the sun's diameter as described.

Image Formation Through a Lens: Find out what happens with more than one pinhole. You can do this by poking an extra hole in your pinhole "camera," or poking an extra hole in a piece of cardboard that you hold between a bright region (window or overhead light) and a dark viewing area. Two holes will produce two images. Poke a third hole and you have three images. Many holes, like a piece of pegboard, produce many images—one for each hole. Here's the neat part: Place a convex lens behind (or in front of) the holes and show how all the images are focused in one place. At the proper distance they neatly overlap to produce a clear and brighter image. So an appropriately-placed lens simply directs a multitude of images atop one another!

You can go further. Choose a bright scene with noticable depth so that perspective plays a role when viewed from close positions (for example, from each eye). Perspective noticeably alters the images through widely-spaced holes. When these different images are combined through a lens, parts of the composite image are fuzzy. A composite image through closely-spaced holes, however, is sharp. This is why photographers take pictures through a small aperture for "depth of field" in their photographs. All parts of an image seen through a small aperture are sharper. When a fuzzy background is desirable, in a flower closeup for example, the aperture of the camera is opened wide. The focal plane is set for sharpness of the flower, and things closer or farther appear fuzzy.

CHECK QUESTION: When the aperture setting of a camera is small, should the exposure time be correspondingly longer or shorter? [The smaller aperture means less light per time, so the film should be exposed to light for a longer time.]

Ray Diagrams: At the board, discuss the key features of a converging lens (text Figure 30-3) and sketch a ray diagram for the case of parallel light along the principle axis. This defines the focal point. Then sketch a diagram for a closer object, such as Figure 30-9. Discuss the rules for ray-diagram construction, page 457, and go over the various cases of text Figure 30-11.

Learning takes place not in seeing diagrams made but rather in making them. Take time to have students make diagrams, preferably for lenses they have in their hands and the objects and images they are witnessing.

After progress has been made with converging lenses, consider making ray diagrams with diverging lenses. Keep in mind, however, that they're more difficult to work with. I suggest avoiding too much emphasis on diverging lenses.

Optical Instruments: As much as possible, show examples of parts of the various optical instruments discussed briefly in the text. Note the simplicity of the text diagrams of these instruments compared to their actual construction. A valuable lesson is to be learned in looking for the simplicity that underlies the seemingly complex.

The Eye: Here you'll be overlapping what students learn about the eye in their life science courses. If they haven't done the blind-spot experiment before, they'll be fascinated with it.

DEMONSTRATION [30-2]: Simulate the human eye with a spherical flask filled with a bit of fluorescent dye. Paint an "iris" on the flask and position the appropriate lens in back of the iris for normal, farsighted, and nearsighted vision. Then show how corrective lenses placed in front of the eye focus the light on the retina.

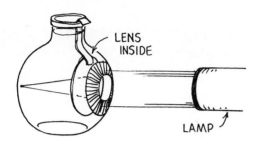

You have probably already discussed why the pupil of the eye looks black. Flashbulb photos often show it to be pink. This is because the light from the flash reflects directly back from the assemblage of blood vessels on the retina. Of interest (and not in the text) is the particularly bright reflection from an animal's eye pupils when illuminated at nighttime. They are so bright they look luminous. This results from the refection of light from a thin membrane located in back of the rods in the animal's eyes. Thus animals are provided with a "second chance" to perceive light that initially misses the rods. This arrangement, common in owls and cats, gives nocturnal predators excellent night vision.

Color Vision (Optional): Discuss the function of the rods and cones in the retina of the eye and how color cannot be perceived in dim light. Point out that colored stars appear white to us, but they show up clearly in color on camera time exposures. (I show a colored slide that I took of the stars, and discuss the curved lines encircling the north star and how long the camera shutter was held open.)

Discuss the fact that we see only motion and no color at the periphery of our vision where rods are located. Prove this fact by conducting the following demonstration.

DEMONSTRATION [30-3]: Stand at a corner of the room and shake brightly colored cards, first turned backward so the color is hidden. Have students look toward the other corner of the room. When they can just barely see the moving cards at the corners of their eyes, turn the cards over to display the color. Try this with different colors. Your students will see the cards as they move, but they will not distinguish the color. This surprising fact amuses the students.

Pupilometrics (Optional): The study of the size of the pupil as a function of attitudes is called pupilometrics (see *Scientific American*, April, 1965). A lot of brain functioning occurs in the eye itself—the eye does some of our "thinking." This thinking is betrayed by the iris, which contracts and expands to regulate the size of the pupil for admitting more or less light as intensity changes. Its changing size is also related to our emotions. If we see, smell, taste, or hear something that is pleasing to us, our pupils automatically increase in size. If we sense something repugnant, they automically contract. Many a card player has betrayed the value of a hand by the size of his pupils. This topic is very interesting. CAUTION: If you discuss this phenomenon, it is important to dispel misconceptions your students may have regarding pupil size. People who normally have small pupils may feel self-conscious about this, and mistakenly feel that small pupils display a negativity of some sort. Also, pupil size decreases with age. Emphasize that it is the CHANGE in pupil size, and not the pupil size itself, that pupilometrics is about.

More Think-and-Explain Questions

1. If you've ever watched a water strider or other insect upon the surface of water, you may have noticed a large shadow cast by the contact point where the thin legs touch the water surface. Then around the shadow is a bright ring. What accounts for this?

   Answer: The "nonwettable" leg of the water strider depresses and curves the surface of the water. This effectively produces a lens that directs light away from its course to form a bright ring around a darker region, which then appears as a shadow. (Interestingly enough, the overall brightness of the shadow and bright ring averaged together is the same whether or not the water is depressed — "conservation of light.")

2. It is often stated that you look at the image of an image when you look through a telescope. What is meant by this?

   Answer: The eyepiece looks at the image of the object cast by the objective lens.

Computational Problem

1. No glass is prefectly transparent. Mainly due to reflections, about 92% of incident light passes through an average sheet of clear glass. The 8% loss is not noticeable through a single sheet. However, through several sheets, the loss is apparent. How much light is transmitted by two sheets?

   Answer: The amount of light transmitted through two sheets of glass is 84.6%. To see this, consider an incident intensity of 100 units. Initially, 92 units are transmitted through the first pane. Then, 92% of this amount is transmitted through the second pane (0.92 of 92 units = 84.6 units, or 84.6% of the incident light).

To use this planning guide work from left to right and top to bottom.

Chapter 31 Planning Guide
• *The bulleted items are key: Be sure to do them!*

Topic	Exploration	Concept Development	Application
Huygens' Principle		• Text 31.1/Lecture	
Diffraction and Interference		• Text 31.2 – 31.4/Lecture Con Dev Pract Pg 31-1 Demo 31-1	
Thin Films	Act 85 (1 period)	• Text 31.5, 31.6/Lecture Demos 31-2, 31-3	Nx-Time Q 31-1
Laser Light		• Text 31.7/Lecture Demo 31-4	
Holograms		• Text 31.8/Lecture	

Video: *None*
Evaluation: Chapter 31 Test

See the Chapter Notes for alternative ways to use all these resources

Objectives

After studying Chapter 31, students will be able to:
• Explain why water waves have curved wave fronts after passing through a narrow opening.
• Describe the conditions necessary for visible diffraction of waves.
• Describe the conditions necessary for visible bright and dark fringes of light caused by interference.
• Explain what causes the bright and dark bands that appear when monochromatic light is reflected from a thin material.
• Explain what causes the colors that shine from soap bubbles or gasoline slicks on a wet surface.
• Distinguish between light from a laser and light from a lamp.
• Distinguish between a hologram and a photograph.

Possible Misconceptions to Correct

• Light travels only in straight lines.
• Light cannot cancel light.

• A laser is an energy source that can put out more energy than it consumes.
• A laser is a high-efficiency device that emits more than just light.
• A hologram is a mystery of science.

Demonstration Equipment

• [31-1] Water-filled ripple tank with wave barriers and a dowel
• [31-2] Index cards with slits cut in them, vertical lamp or fluorescent lamp separated into three colored segments using red, clear, and blue plastic
• [31-3] Laser, piece of glass with irregular surface (shower door glass or cut crystal glassware), and music (optional)
• [31-4] Polarized slides 5 cm × 5 cm with crumpled cellophane or shiny cellophane tape, and slide projector
• [31-5] Laser and a means to show laser lissajous patterns

Introduction

This chapter is best taught with a ripple tank. If you do not have a commercial tank that shows detail on a screen, a large pan will serve the purpose. As already suggested in Chapter 29, use an inexpensive one-piece frame of clear plastic a little more than an inch deep. Put water in it and mount it on an overhead projector. Put strips of screen mesh around the edges to reduce unwanted reflected waves. Use a large-diameter wooden dowel to generate waves. A gentle roll forward followed by a quick roll backward produces a nice single pulse. You control the frequency of additional pulses. If you wish to make custom-designed shapes to show diffraction, either an acrylic plastic material called Plexiglass® or a polycarbonate material called Lexan® can be used.

If you use music with the three "light shows" suggested here, you'll impress your students with the beauty of physics. These light shows can be among the high points of your course, so do not show all three of them in a single lecture.

If you view interference colors in soap bubbles, you might mention that the thin film of the bubble is about the thinnest film seen with the unaided eye.

Suggested Lecture

Start by acknowledging Huygens' principle: Not only can waves combine to form bigger waves, but waves can also be considered to be made up of smaller waves.

Diffraction: Introduce this topic by using the following demonstrations.

DEMONSTRATION [31-1]: Show the diffraction of water waves in a ripple tank. Arrange objects in the tank to diffract waves through a variety of slit sizes. Water waves are your example, but these wave properties apply also to sound waves, light waves, and ALL kinds of waves.

DEMONSTRATION [31-2]: After discussing diffraction, pass around some index cards with razor slits in them. Illuminate a vertical or overhead fluorescent lamp that has been separated into three segments by colored plastic; red, clear, and blue. Have your students view the diffraction of each color segment through the slit, or they can view them through a slit provided by their own fingers. Note the various fringe spacings of different colors.

Interference: This topic was introduced in Chapter 25 and applied to sound in Chapter 26.

DEMONSTRATION [31-3]: With lights out, shine a laser light through a piece of irregular glass (shower door glass, sugar bowl cover, or cut crystal glassware) to display beautiful interference patterns on the wall. This is especially effective if you make slight movements of the glass in rhythm with music. Your students will not forget this demonstration!

Thin-Film Interference: After your students do the lab activity *Rainbows Without Rain*, shine the light from a sodium lamp (or laser light) diffused through a diverging lens onto a pair of glass plates, as shown in text Figure 31-17. Relate the interference fringes to those explained in Figure 31-15. It is important that Figure 31-15 be understood, so spend some time discussing it.

CHECK QUESTION: What is the similarity between Figure 31-15 on text page 478 and Figure 26-14 on page 389? [They both illustrate the same phenomenon of interference, the inference of one is of sound and the other is of light. A big difference is the wavelengths involved. The wavelength of light is much smaller — hence the narrower fringes.]

Iridescence from Thin Films: Talk about the iridescent colors that appear to change with position in the feathers of some birds (peacock, pigeon, starling). Discuss the spectrum of colors seen in soap bubbles and in certain seashells. State that the colors are produced by interference. Illustrate the colors reflected from gasoline on a wet street by sketching Figure 31-20 on the board.

CHECK QUESTION: Why are interference colors produced by gasoline spilled on a dry surface? [Two smooth reflective surfaces are necessary for interference colors to occur. Without the smooth water beneath the film of gasoline, the lower surface of the gasoline would not be a smooth reflector.]

Emphasize why two surfaces are needed to produce interference colors and why the film should be thin. Explain that the recombination of "split" waves cannot occur when the reflected rays are widely displaced.

A nice example of interference not given in the text is that of the bluish tint of coated lenses. Lenses are coated to destroy doubly-reflected light in a camera lens. When light that doubly reflects in a lens arrives at the film, it is out of focus. All the wavelengths of light that are doubly reflected can't be destroyed by interference with a single thin film. But a film thickness one quarter of the wavelength of yellow light will cancel the most predominant color in sunlight — yellow. So the sunlight you see reflected from a coated lens is deficient in yellow and appears blue.

DEMONSTRATION [31-4]: The vivid colors that emerge from cellophane between crossed polarized lenses make a spectacular demonstration. Have students make up some slides (5 cm × 5 cm) of cut and crinkled cellophane mounted on polarized material. Place one in a slide projector and rotate a sheet of polarized material in front of the projecting lens, so that a changing montage of colors is displayed on the screen. Also include a show of color slides of the interference colors seen in the everyday environment, as well as of microscopic crystals. This is more effective with two projectors with hand dissolving from image to image on the screen. Do this in rhythm to some music and you'll have an unforgettable lecture demonstration!

Laser Light: Correct the misconception that a laser is a powerful emitter of something other than light and that it is an efficient light source. Your classroom laser is typically less than 1% efficient! If you showed the spectrum of neon in Chapter 28, state that the frequency of light emitted from a helium-neon laser is just one of those many neon spectral lines.

Here's some additional helium-neon laser information. In the laser, the low pressure mixture of 85% helium and 15% neon is subjected to a high voltage. This energizes (excites) the helium. Before the helium radiates light, it collides with neon atoms in the ground state and transfers energy to them. The amount of energy is just sufficient to excite neon to an otherwise difficult-to-come-by, metastable state that is very close to the energy of the excited helium. The process continues and the population of excited neon atoms outnumbers neon in the ground state. This inverted population is, in effect, waiting to radiate its energy. When some neon atoms emit light, the radiation passes other excited neon atoms and triggers their de-excitation, exactly in phase with the stimulating radiation. Light parallel to the tube bounces from specially coated mirrors, and the process cascades to produce a beam of coherent light. (To reduce information overload, note that the excitation model of light emission is not treated in the text, and it is only briefly mentioned in the footnote on page 427 in Chapter 28.)

DEMONSTRATION [31-5]: Give a laser show. Sprinkle chalk dust or smoke in the laser beam. Show diffraction through a thin slit. An unforgettable presentation is directing a laser beam on a mirror which has been fastened to a rubber membrane stretched over a radio loudspeaker. Do this to music and cover the darkened walls with a display of dancing lissajous patterns.

Holograms: Here's an effective sequence for explaining holograms. With the aid of text Figure 31-15, develop the idea of interference for multiple slits, which constitute the diffraction grating. With a large diffraction grating (the size of a full sheet of typing paper), show the spectral lines of a gas discharge tube. Emphasize that physical lines do not really exist where they appear to be. The lines are virtual images of the glowing tube (just as they would be images of slits, if a slit were being used). Now that students have a fairly good idea of how these images of spectral lines are produced by the diffraction grating, show them a really sophisticated diffraction grating. Show them one of microscopic swirls of lines in two dimensions — a hologram illuminated with a laser.

More Think-and-Explain Questions

1. A monochromatic light illuminates two closely-spaced thin slits and produces an interference pattern on the wall. How will the distance between the fringes in the pattern be different for blue light than they are for red light?

   Answer: The fringes will be spaced closer together if the pattern is made of shorter-wavelength blue light. The longer-wavelength red light will produce fringes that are farther apart. (Investigation of text Figure 31-15 should make this clear. Note that when the wavelength is shorter, the light and dark regions on the screen are closer together.)

2. Why do the iridescent colors seen in some seashells, such as the inside of abalone shells, change when the shell is viewed from a different position?

   Answer: The optical path of light from upper and lower reflecting surfaces changes when position of the shell is changed. Thus different colors are seen when the shell is held at varying angles.

3. If you are viewing a hologram and you close one eye, will you still perceive depth? Explain.

   Answer: Two eyes are required to perceive depth by parallax, whether in a hologram or otherwise. If depth is perceived by other cues, such as relative sizes and relative brightness of objects, then one eye is sufficient.

32 Electrostatics

To use this planning guide work from left to right and top to bottom.

Chapter 32 Planning Guide

• The bulleted items are key: Be sure to do them!

Topic	Exploration	Concept Development	Application
Electric Forces and Charges		• Text 32.1/Lecture • Demo 32-1	
Conservation of Charge		• Text 32.2/Lecture Demos 32-2 – 32-5	
Coulomb's Law		• Text 32.3/Lecture • Con Dev Pract Pg 32-1	Nx-Time Q 32-1
Conductors and Insulators		• Text 32.4/Lecture	
Charging	Act 86 (1 period)	• Text 32.5, 32.6/Lecture • Con Dev Pract Pg 32-2	
Charge Polarization		• Text 32.7/Lecture Demos 32-6 – 32-8	

Video: *Electrostatics*
Evaluation: Chapter 32 Test

See the Chapter Notes for alternative ways to use all these resources

Objectives

After studying Chapter 32, students will be able to:
- Describe electrical forces between objects.
- Explain, from the point of view of electron transfer, how an object becomes positively charged or negatively charged and relate this to the object's net charge.
- Describe the relationship of the electrical force between two charged objects, the charge of each object, and the distance between the charges.
- Compare the strength of electrical force and that of gravitational force between two charged objects.
- Distinguish between a conductor and an insulator.
- Describe how an insulator can be charged by friction.
- Describe how a conductor can be charged by contact.
- Describe how a conductor can be charged without contact.
- Describe how an insulator can be charged by charge polarization.

Possible Misconceptions to Correct

- The study of electricity is incomprehensible.
- Electric charges occur in some materials and not in others.
- Friction is a necessary factor in charging an object.
- Lightning rods are designed to attract lightning.

Demonstration Equipment

- [32-1] Fur, silk, rubber rod, glass or plastic rod, suspended pith ball
- [32-2] Electrophorus; metal plate with insulating handle, piece of Plexiglass® acrylic plastic, regular plastic, or equivalent nonconductor
- [32-3] Whimshurst machine (electrostatic generator), a sharp metal point, an alligator clip
- [32-4] Vane of metal points that rotate on a needle pivot when charged
- [32-5] Van de Graaff generator, aluminum pie pans, puffed rice or puffed wheat, lamp tube
- [32-6] Rubber balloon
- [32-7] A piece of wooden 2 × 4, about a meter long, balanced upon a watch glass or metal spoon so it will rotate
- [32-8] Charged rubber rod and stream of water

Introduction

Many facts about electricity are generally misunderstood not only by the general public, but also by people involved in technology.

The study of electricity begins with electrostatics, which is best introduced as a series of coordinated demonstrations. After demonstrating charging with fur and rubber rods to show electrostatic attraction and repulsion (Coulomb's law), do the following demonstrations in order: (1) the electrophorus (a metal plate charged by induction by a sheet of Plexiglass which has been charged with fur. You can substitute a pizza pan and a phonograph record), (2) the Whimshurst machine (electrostatic generator), and (3) the Van de Graaff generator.

The lab activity, *Static Cling*, can follow your introductory lecture, and it can replace lectures on Sections 32.6 *Charging by Induction* and 32.7 *Charge Polarization*. There are no follow-up experiments for this chapter, nor are there any for the next chapter. For practical purposes, this and the following chapter are theoretical background for the study of electric current in Chapter 34 and circuits in Chapter 35. Lab experiments are featured for those chapters. So this chapter is prerequisite to the following chapters in Unit 5.

Suggested Lecture

Begin by showing the following familiar charging demonstration.

DEMONSTRATION [32-1]: Rub the fur on a rubber rod to charge both objects. Deposit some of the rod's acquired negative charge on a suspended pith ball. Recharge the rod and bring it near the pith ball. The observed repulsion shows the first part of the fundamental rule of electricity: Like charges repel. Then rub silk and a glass rod together. When the positively-charged glass rod is brought near the pith ball, attraction is observed. This illustrates the second part of the fundamental rule: Unlike charges attract.

Cite the enormously greater force of electricity compared to force of gravity. Only a small amount of net charge produces noticeable effects. The gravitational interaction between the same objects is completely negligible by contrast. Explain what it means to say an object is electrically charged and discuss the conservation of charge. Compare this to taking bricks from a road and putting them on the sidewalk.

Charging something can be compared to removing bricks from a road and putting them on a sidewalk. There are exactly as many "holes" in the road as there are bricks on the sidewalk.

DEMONSTRATION [32-2]: Show the electrophorus — a metal plate with an insulating handle and a piece of Plexiglass, regular plastic, or some equivalent nonconductor. Slap the fur on the Plexiglass several times to charge it. Rest the metal plate on the charged Plexiglass. Be careful not to ground it. Remove the plate and place it near the charged pith ball. No interaction is observed (providing you don't produce air currents by your action). Then again place the plate on the Plexiglass, only this time touch the plate with your finger. You are grounding the plate by producing a path to ground for the electrons that have been pushed to the top surface of the plate. The plate is now positive, as you demonstrate by bringing it near the pith ball, which is attracted to the plate. Touch the plate to the end of a neon or fluorescent tube and show the flash of light that is produced. Explain the process of grounding when you touch your finger to the plate.

DEMONSTRATION [32-3]: Demonstrate the Whimhurst machine and explain its similarity to the electrophorus. Actually, the Whimhurst machine is a rotating electrophorus! Show sparks jumping between the spheres of the machine and discuss the sizes (radii of curvature) of the spheres in terms of their capacity for storing charge. (The amount of charge that can be stored before discharge to the air is proportional to the radius of the sphere. Large spheres can hold large quantities of charge, and spheres of small radii can hold only small quantities of charge.) Fasten a sharp metal point, which has a tiny radius of curvature and hence a tiny charge storing capacity, to one of the Whimshurst spheres and demonstrate the leakage of charge.

If you wish to expand on the idea of charge leakage from a point, simplify it. Explain that on the surface of an electrically-charged flat metal plate, every charge is mutually repelled by every other charge. If the surface is curved, charges on one part of the plate will not interact with charges on some distant part of the plate. Because of the shielding effect of the metal, they are out of the line of sight of one another. Hence for the same amount of work or potential, a greater number of charges can be placed on a curved surface than can be placed on a flat surface. The more pronounced the curvature is, the more charge can be stored. To carry this idea further, consider a charged needle. Under mutual repulsion, charges gather to the region of greatest curvature, which is the point. Although all parts of the needle are charged to the same electric potential, the charge density is greatest at the point. The electric field intensity about the needle, on the other hand, is greatest about the point. It is usually great enough to ionize the surrounding air and to provide a conducting path from the charge concentration. Hence, electric charge readily gathers at points and readily leaks from points.

DEMONSTRATION [32-4]: Illustrate leakage and the reaction force (ion propulsion) with a set of metal points arranged so that they rotate when charged.

Lightning Rods: Discuss lightning rods and explain how induced charges gather at the points and leak off into the air. Explain that the little spheres on the tip of automobile radio antennas prevent leakage as well as static on the radio.

Show how the bottoms of negatively-charged clouds and the resulting induced positive charge on the earth's surface are similar to the electrophorus held upside down. Compare the charged Plexiglass plate to the clouds and compare the metal plate to the earth. After sketching the charges for the clouds and the earth on the chalkboard (see text Figure 32-10), be sure to hold the inverted electrophorus pieces against your drawing on the board to show their respective places. Discuss the lightning rod as a safeguard against lightning while showing the similar function of the metal point attached to the Whimhurst machine. (Notice that one idea flows to the next in this sequence. This is very important, since the ideas of electricity are usually difficult to grasp the first time. Be sure to move carefully through this sequence of demonstrations and their explanations.)

Benjamin Franklin was the first to show that a kite flown during a thunderstorm collects electric charges from the air. Hairs on the kite string stood apart, implying that lightning was a huge electric spark. Franklin's kite, by the way, was not struck by lightning. If it had been, he would probably not have been around to report his experience. Others who repeated his experiment were not so fortunate.

After establishing the idea that charge capacity depends on the area and curvature of the conductor being charged, introduce the Van de Graaff generator.

DEMONSTRATION [32-5]: Van de Graaff-generator time: Set a cup on top of the uncharged dome and fill it with puffed rice or puffed wheat. Your students will like the fountain that follows when you charge the cup. You can also place a stack of aluminum pie pans on the dome. When the dome is charged, the pans fly off one at a time, showing electrostatic repulsion. Quite impressive!

Introduce the idea of the electric field at this time. Call attention to the space surrounding the charged Van de Graaff dome and state that a force field is kind of extended aura that surrounds something and spreads its influence to the surroundings. This is be the focus of the next chapter.

Charge Polarization: Define charge polarization by explaining Figures 32-11 through 32-14 in the text.

DEMONSTRATION [32-6]: Rub an inflated balloon on your hair and stick on the wall. Follow this up with a sketch of text Figure 32-13.

DEMONSTRATION [32-7]: Place a charged rod near the ends of a section of wooden 2 × 4 more than a meter long; balance and rotate the wood sideways at its midpoint by placing it on the bottom of a metal spoon. You can easily set the massive piece of dry wood in motion.

DEMONSTRATION [32-8]: Place a charged rod near a thin stream of falling water. The stream of water will be attracted to the charged rod. The fact that it is attracted to either a positive or a negative rod shows that the stream itself is not charged. The explanation for this attraction is that the polar molecules of water swing into alignment like compass needles aligning with a magnet.

More Think-and-Explain Questions

1. The two leaves of an electroscope repel each other and stand out at an angle. What balances the electrical force to keep the leaves from standing out farther?

   Answer: The weight of the leaves under the influence of gravity influences the angle.

2. If a positive test charge is placed between a pair of oppositely-charged plates, which way will the test charge move?

   Answer: It will be attracted to and move toward the negative plate. Put another way, it will be repelled by the positive plate and move toward the negative plate. (In the next chapter we will see that it will move in the direction of the electric field, from positive to negative.)

Computational Problems

1. What is the electrical repulsion between two objects that each carry 1 C of charge and are 1 meter apart?

   Answer: $F = k\ qq/d^2 = 9 \times 10^{-9}$ Nm^2/C^2 $(1\ C)^2/(1\ m)^2 = 9 \times 10^9$ N. That's nine billion newtons! Obviously we don't experience charges of 1.0 C in our everyday environment.

2. Compute the gravitational force between two electrons ($m = 9.1 \times 10^{-31}$ kg) separated by 1 cm. Compare your answer to the electrical repulsion.

   Answer: $F_g = G\ mm/d^2 = 6.67 \times 10^{-11}$ $Nm^2/kg^2(9.1 \times 10^{-31}kg)^2/(0.01\ m)^2 = 5.5 \times 10^{-67}$ N. For electric repulsion, $F = k\ qq/d^2 = 9 \times 10^9$ Nm^2/C^2 $(1.6 \times 10^{-19}\ C)^2/(0.01\ m)^2 = 2.3 \times 10^{-24}$ N. So we see $F_e/F_g = 2.3 \times 10^{-42}\ N/5.5 \times 10^{-67}$ N $= 4.2 \times 10^{42}$. So the electric force is more than 10^{42} times greater than the gravitational force is. No wonder that electronics people neglect the effects of gravity between charges! (Note that this ratio is different than the ratio of similar forces shown on the text pages 497 and 498, because the relative masses and hence gravitational forces are different than the text example.)

3. How much force acts on an electron placed in an electric field of strength 1000 N/C? At what rate will the electron accelerate in this field?

   Answer: $F_e = Eq = (1000\ N/C)(1.6 \times 10^{-19}C) = 1.6 \times 10^{-16}$ N.

   $a = F_e/m = 1.6 \times 10^{-16}\ N/9.1 \times 10^{-31}kg = 1.8 \times 10^{14}$ m/s^2. Compared with the acceleration due to gravity at the earth's surface, this is a very high acceleration. It is due to the incredibly small mass of the electron.

4. The force of attraction between a certain negative charge and a positive charge is 100 N. What will be the force of attraction if the distance between the charges is (a) doubled? (b) tripled? (c) quadrupled?

   Answer: By the inverse-square law, the force of attraction will be: (a) 1/4, (b) 1/9, (c) 1/16.

33   Electric Fields and Potential

To use this planning guide work from left to right and top to bottom.

Chapter 33 Planning Guide
• The bulleted items are key: Be sure to do them!

Topic	Exploration	Concept Development	Application
Electric Fields		• Text 33.1/Lecture	
Field Lines		• Text 33.2/Lecture Con Dev Pract Pg 33-1	
Shielding		• Text 33.3/Lecture	Nx-Time Q 33-1
Electric Potential Energy		• Text 33.4/Lecture Demo 33-1	
Electric Potential		• Text 33.5/Lecture Con Dev Pract Pg 33-2	
Energy Storage (Capacitors)	Act 87 (1 period)	• Text 33.6/Lecture	
Van de Graaff Generator		Text 33.7/Lecture	Nx-Time Q 33-2
Video: *None* **Evaluation:** Chapter 33 Test			

See the Chapter Notes for alternative ways to use all these resources

Objectives

After studying Chapter 33 students will be able to:

- Describe how the intensity of an electric field at two different points can be compared.
- Describe how the direction of an electric field at a point is determined.
- Relate the spacing of electric field lines to the intensity of the electric field.
- Describe the conditions under which something can be completely shielded from an electric field.
- Explain why a charged object in an electric field is considered to have electric potential energy.
- Distinguish between electric potential energy and electric potential.
- Describe the purpose of a Van de Graaff generator.

Possible Misconceptions to Correct

- Electric potential energy and electric potential are the same thing.
- A capacitor is a source of electric energy.
- High voltage is dangerous under any conditions.
- The voltage produced by rubbing a balloon on one's hair is low compared to the voltage of electric circuits in the household.

Demonstration Equipment

- [33-1] Van de Graaff generator, lamp tube

Introduction

This is a theoretical chapter and mastering it is a challenge. It is recommended that the effort be spent instead on the subsequent chapters that are supported with hands-on activities. You should move quickly through this chapter and treat it as an overview, rather than as material to be mastered.

Understanding the vector nature of fields is enhanced with the Practice Page for this chapter, though its use depends on prior treatment of vectors (see Chapter 6).

This edition includes a brief study of capacitors. Capacitors and light bulbs can be used effectively to help students visualize the concepts of electricity. Some neat ideas are incorporated in the new lab experiment, *Brown Out*.

Suggested Lecture

Electric Fields: Begin by stating the common usage of the word field: An area of ground perhaps planted in grasses of different heights. The fields of physics, however, include gravity fields, magnetic fields, and electric fields and are the subject of this chapter.

Compare electric fields to gravity fields. They are similar in that both surround an agent; charge for electric and mass for gravity. They both obey the inverse-square law. The chief difference is that a gravity field only attracts, while an electric field can both attract and repel.

Electric Field Lines: Cite the vector nature of a force field and describe the lines of force (text Figures 33-3 and 33-4). These lines are nicely visible in the photos of Figure 33-5.

CHECK QUESTION: If a tiny test charge were dropped in the oil bath shown in Figure 33-5, in what direction would it move? [Along the same directions as the bits of thread — away from the conductor of opposite charge and toward the conductor of same charge.]

Shielding: Call attention to photo d of text Figure 33-5 which shows that the threads have no directional properties inside the charged cylinder. This shows that the electric field is shielded by the metal. The dramatic photo of the car being struck by lightning in text Figure 33-6 also illustrates that the electric field inside a conductor is normally zero, no matter what is happening outside. The explanation is only hinted at in the text in Figure 33-7. You can leave the discussion at that and move on to new material.

Should some of your students wish more information, you can go a step beyond the test charge in the middle of the sphere (text Figure 33-7). Consider the test charge off-center, twice as far from region A as region B, as shown in the art below. The dotted lines represent a sample cone of action, subtending both A and B. Region A has twice the diameter, four times the area, and four times the charge as region B does. Four times the charge, however, is at twice the distance, which by the inverse-square law will have one-fourth the effect. So the greater charge is balanced by the correspondingly greater distance. This will be the case for all points inside the conductor. The conductor need not be a sphere, as is shown by the shapes in text Figure 33-8.

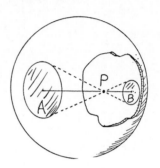

From another viewpoint: If the field inside a solid or hollow conductor were not zero, then the free charges inside would move. Would the movement continue forever? [No, the charges would finally move to positions of equilibrium.] In equilibrium, the effects of the charges on each other would be mutually balanced and complete cancellation of fields everywhere inside the conductor would occur. This is what happens — not gradually, but suddenly!

Electric Potential Energy and Electric Potential: Fields are called *force fields* because forces are exerted on the bodies within them. They could properly be called *energy fields*, because energy is stored in a field. Compare the gravitational potential energy of the lifted mass in text Figure 33-10 with the "lifted" positive charge. Electric potential energy is analogous to gravitational potential energy. We speak about the electric potential energy that charged bodies have in a field, but more often we speak of the potential energy *compared to the amount of charge* — electric potential. Consider the electric field surrounding a charged Van de Graaff generator. Explain that the field energy is greatest close to the charged dome and is weaker with increased distance. Likewise, the electric potential energy of a charge is greatest in the region near the dome and weaker with increased distance away from the dome. Note that your explanation begins with the electric field, goes to the electric potential energy, and then moves to the electric potential. These are all abstract concepts that may not be readily grasped by your students. The following demonstration might help.

DEMONSTATION [32-1]: Hold a fluorescent lamp tube at the end and place it in the field of a charged Van de Graaff generator. Show that it lights up when one end of the tube is closer to the dome than the other end is. Introduce the idea of potential difference. The end of the tube closest to the generator is at a higher electric potential than the other end is. There is a potential difference across the ends of the tube so it lights up. Then show that when both ends of the fluorescent tube are equidistant from the charged dome (at the same potential), light emission ceases. This phenomenon sets the groundwork for your next lecture on electric current.

Go lightly on units of measurement. Some students will learn that $N/C = V/m$ and $J/C = V$ without your describing it in class. The art of teaching introductory physics is using a broad approach that gives an overview. Detail is something that will come later to those who need it. If you do have a student or two who want details now, supply them on a one-to-one basis. Don't be afraid to skip over detail when it's counterproductive!

Capacitors: A charged capacitor is a useful device not only for storing electric energy, but also for studying electrical properties. Although primarily used in ac circuits, capacitors can be a quasi-source of electric energy and serve as a substitute for a battery in a hands-on circuit experience. This is hands-on time if you are doing the experiment *Brown Out* described in the Lab Manual. Capacitors are treated again in Chapter 34.

More Think-and-Explain Questions

1. Measurements show that there is an electric field surrounding the earth. Its magnitude is about 100 N/C at the earth's surface and the field points inward toward the earth's center. From this information, can you state whether the charge of the earth is negative or positive?

   Answer: The earth is negatively charged. If it were positive, the field would point outward.

2. Would you feel any electrical effects if you were inside the charged sphere of a Van de Graaff generator? Why or why not?

   Answer: You would feel no electrical effects inside any charged conducting body. Because of mutual repulsion, all charge resides on the outside surface of the conductor. The distribution of charge is such that complete cancellation of the interior electric field occurs. The electric field inside any conductor under static conditions is zero, whether the conductor is charged or not. (If the electric field were not zero, then conduction electrons would move in response to the field until electrical equilibrium was established — a zero electric field.)

Computational Problems

1. If 10 joules of work must be done to move 2.0 C of charge between two plates, what is the potential difference between the plates?

   Answer: $V = E/q = 10$ J/2.0 C = 5 volts.

2. How much work is required to move 5.0×10^{-12} C of charge between two plates if the potential difference between the plates is 100 volts?

   Answer: $W = E = Vq = (100$ V$)(5.0 \times 10^{-12}$C$)$ = $(100$ J/C$)(5.0 \times 10^{-12}$ C$) = 5 \times 10^{-10}$ J.

To use this planning guide work from left to right and top to bottom.

Chapter 34 Planning Guide

• *The bulleted items are key: Be sure to do them!*

Topic	Exploration	Concept Development	Application
Flow of Charge	• Act 88 (1 period)	• Text 34.1/Lecture • Demo 34-1	
Electric Current		• Text 34.2/Lecture	
Voltage Sources		• Text 34.3/Lecture	Nx-Time Q 34-1
Electrical Resistance		• Text 34.4/Lecture	
Ohm's Law and Electric Shock		• Text 34.5, 34.6/Lecture • Con Dev Pract Pg 34-1 • Con Dev Pract Pg 34-2	Exp 89 (>1 period)
AC and DC		• Text 34.7/Lecture	
AC - DC Conversion		• Text 34.8/Lecture Demo 34-2	
Electron Speed		• Text 34.9/Lecture	
Electron Source		• Text 34.10/Lecture	
Electric Power		• Text 34.11/Lecture Con Dev Pract Pg 34-3	

Video: *Electric Current*
Evaluation: Chapter 34 Test

See the Chapter Notes for alternative ways to use all these resources

Objectives

After studying Chapter 34, students will be able to:

• Describe the conditions necessary for electric charge to flow.

• Describe what is happening inside a current-carrying wire and explain why there is no net charge in the wire.

• Give examples of voltage sources that can maintain a potential difference in an electric circuit.

• Describe the factors that determine the resistance of a wire.

• Relate the current in a circuit to the resistance of the circuit and the voltage across it.

• Explain why wet skin increases the likelihood of receiving a damaging electric shock when a faulty electrical device is touched.

• Distinguish between direct current and alternating current.

• Distinguish between drift speed of conduction electrons in a current-carrying wire and signal speed.

• Relate the power used by an electrical device to its current and voltage.

Possible Misconceptions to Correct

- A current-carrying wire is electrically charged.
- Electric current *is* a fluid of some kind.
- Electric current flows out of and into a battery, rather than flowing through a battery.
- Voltage flows through a circuit, instead of being impressed across a circuit.
- Power companies deliver electrons, rather than energy, from a power plant to consumers.
- Electrons travel at about the speed of light in a dc circuit.

- A voltage or power source supplies electrons to a circuit to which it is connected.
- Mechanical power is the time-rate of doing work, whereas electrical power the product of voltage and current.

Demonstration Equipment

- [34-1] Simple circuit of three lamps and a battery of any kind
- [34-2] A light-emitting diode (LED) and a battery

Introduction

The intent of this chapter is to build a good understanding of current electricity and to dispel some of the popular misconceptions about electricity. If you are teaching only one chapter on electricity, this should be it.

Ohm's law is the core of the chapter. No distinction is made between resistors that change their value and those that don't when voltage changes. We simply say that whatever the conductor, the current produced by an impressed voltage is equal to the ratio of that voltage to the particular resistance. In this sense, Ohm's law is conceptually useful.

As a practical matter, Ohm's law is useful for predicting values only if the resistance of the device does not change with changes in voltage or current, and usually where heating does not appreciably affect resistance. Devices that keep the same resistance for a wide range of voltages are said to be *ohmic*. Ohm's law is useful for predicting values of ohmic materials.

A new topic in this edition is the diode, a tiny electronic device that allows current flow in a single direction and which is effective in converting ac to dc. When combined with a capacitor, the pulsing ac is smoothed to dc.

Suggested Lecture

Begin by reviewing the lighting of a fluorescent lamp or a neon discharge tube with the Van de Graaff generator from your last lecture. Explain this in terms of current being directly proportional to voltage. Relate voltage to the idea of electric pressure. Emphasize that a *difference* in electric potential must exist, or if you prefer, use voltage *difference*. Cite examples, such as birds sitting on bare high-potential wires and the inadvisability of using electric appliances in the bathtub. Discuss electric shock and explain that electricians put one hand behind their back when probing questionable circuits, so current will not pass between their arms and through the heart. Discuss how being electrified produces muscle contractions that can result in not being able to let go of hot wires or in being thrown to the ground.

Ohm's Law: Introduce the idea of electrical resistance to complete the discussion of Ohm's law. Explain that resistance is dependent on the material that the electricity passes through — its cross-sectional area, its length, and its temperature. Compare the resistance of various materials and of various thicknesses of wire of the same metal. Call attention to the ceramic supports for the wires on high-voltage power lines and the rubber insulation that separates the pair of wires in a common lamp cord, and briefly discuss their function.

> DEMONSTRATION [35-1]: Show a very simple series circuit consisting of two or three lamps (bulbs) that are not bright enough to make viewing uncomfortable. Relate the terms in Ohm's law to charge flow, voltage source, and the resistance of the lamp filaments. Interchange lamps of low and high resistance to relate resistance to lamp brightness.

Electric Shock: In your discussion of electric shock, call attention to the Ohm's law formula on the board. First point out the potential (voltage) difference and then the resistance. The human body, like any conductor, conducts current in accordance with Ohm's law. Distinguish between the effect (current) and the cause (voltage). Cite the variability of resistance of the human body under different conditions, for example, dry versus wet.

Current Speed: To impart the idea of dc current in a circuit, do the following. Ask the class to imagine a long column of marchers at the front of the room, all standing at rest close together. (Or instead of an imaginary column, ask for a few volunteers and DO it.) Walk to the end of the column and shove the last member. Ask the class to notice the resulting impulse that travels along the line until the first marcher is jostled against the wall. Then ask if this is a good analogy for how electrons travel in a wire. (The answer is no. This is a good analogy for how sound travels, but not how electrons travel.) Cite how slowly the disturbance traveled through the marchers and how slowly sound travels, when compared to light or electricity. Again call attention to the column of marchers and walk to the far end and call out, forward march! As soon as the command reaches each individual, each steps forward. The marcher at the beginning of the column steps forward immediately. State that this is an analogy for dc electricity. Except for the brief time it takes for the electric *fields* of electrons to travel through the wire (nearly the speed of light), electrons at the far end of the circuit respond immediately. State that the speed at which the "Forward march" command traveled is altogether different from how fast each marcher moved upon receiving that command. Also, the velocity of the electric signal (nearly the speed of light) is quite a bit different than the drift velocity (a snail's pace) of electrons in a circuit.

> CHECK QUESTION: When you turn the ignition key to start your car, electrons migrate from the negative battery terminal through the electric network to the starter motor and back to the positive battery terminal. About how much time is required for an electron to leave the negative terminal and go through the circuit and return? Less than a millisecond? Less than a second? About a second or two? Or about a day? [The correct answer is about a day.]

AC and DC: Discuss the differences between ac and dc. A hydrodynamic analogy for ac is useful. Visualize powering the agitator in a washing machine with water power. A pair of clear plastic pipes connected to a paddle wheel at the bottom of the agitator is fashioned so that water sloshes to and fro in the pipes and rotates the agitator back and forth. Visualize the free ends of the plastic pipe connected to a special socket in the wall. The socket is powered by the power utility. The socket supplies no water, but rather it consists of a couple of pistons that exert a pumping action, one out and the other in, then vice versa, in rapid alternation. When the ends of the pipe containing water are connected to the pistons, the water in the pipes is made to slosh back and forth, thus power is delivered to the washing machine. There is an important point to note here. The power

company supplies no water, just as the power utilities supply no electrons. The greater the load on the agitator, the more energy the power company must deliver to the action of the alternating pistons. This analogy affords a visual model for household current, especially with the transparent plastic pipes.

Converting AC to DC: Batteries produce dc. And battery-driven devices can also be powered with ac, when it is converted to dc. This is accomplished with a diode, a tiny electronic device that allows the passage of charge in only one direction. Half a cycle of ac passes through a diode which is off half the time to produce a rough dc (see text Figure 34-12). Smoothing is accomplished with a capacitor.

> DEMONSTRATION [34-2]: Place the leads of a light-emitting diode (LED) across the terminals of a battery and show how it lights only when the polarity is matched.

Electron Source: Ask for an estimate of the number of electrons pumped by the local power plant into homes and industries in the past year. [Zero!] Then stress the idea that power plants do not sell electrons — they sell energy. You supply the electrons!

Power: Briefly review the definition of power (energy/time) and its units, 1 joule/second = 1 watt. For electrical energy, this is a small unit of power, so it is common to discuss 1000 joules/second = 1 kilowatt. Explain how the unit joule can also be expressed as the unit kilowatt-hour (footnote, page 532) by using an actual electric bill to make your point.

More Think-and-Explain Questions

1. The text stresses that electrons travel very slowly in a dc electric circuit. Why, then, do the lights go off immediately when you turn off a light switch?

   Answer: The electric field goes through the circuit at the speed of light (somewhat less than the speed of light in a vacuum). When the field is cut off, the flow of electrons ceases.

2. If a current that is one tenth to two tenths of an ampere flows into one hand and out the other, you will probably be electrocuted. But, if the same current passes between your hand and your elbow of the same arm, you will survive. Explain.

   Answer: In the first case, the current probably passes through your heart. In the second case, it passes only through your forearm.

3. Energy is put *into* electricity by pumping electrons from a low voltage to a high voltage. How do we get energy *out of* electricity?

   Answer: Energy output is obtained when

electrons flow from a high voltage to a lower voltage. In so doing, the energy (from a battery or generator) is usually transformed to other forms and/or to other places (heat in toasters, mechanical motion in electrical toothbrushes, sound in radios).

4. Which will do more damage, plugging a 120-volt appliance into a 240-volt circuit or plugging a 220-volt appliance into a 120-volt circuit? Explain.

Answer: Damage generally occurs from excess heating caused by too much current through an appliance. For an appliance that converts electrical energy directly to heat, this happens when excess voltage is applied. So don't connect a 120-volt appliance to a 240-volt circuit. Interestingly enough, if the appliance is an electric motor, then applying too *little* voltage can result in the motor overheating and burning up the motor windings. (This occurs when the motor spins at a low speed and the reverse *generator effect* (back emf) is small and allows too great a current to flow in the motor.) Don't hook up a 240-volt motor-driven appliance to 120 volts. To be safe, use the recommended voltage for the appliance.

5. Is a current-carrying wire electrically charged?

Answer: The net charge in a wire, current or not, is normally zero. That's because the number of electrons is ordinarily offset by an equal number of protons in the atomic lattice. Consequently, current (the net flow of charge) and the charge itself are not the same thing. Many people think that saying a wire carries current is the same thing as saying a wire is charged. Perhaps they mean the wire is "energized" with the uniform motion of charges, which is not the same thing. A wire that carries a current is typically not electrically charged and won't affect an electroscope. If current consists of a flow of electrons in the absence of an equal number of protons, say a beam of electrons in a vacuum, then the beam is charged.

6. Which has the thicker lamp filament, a 60-watt bulb or a 100-watt bulb?

Answer: The 100-watt bulb does. This follows from the fact that more current flows in a 100-watt bulb. This is consistent with the relationship *power = current × voltage*. More current for the same voltage means less resistance, so a 100-watt bulb has less resistance than a 60-watt bulb. Less resistance for the same material means a thicker (or shorter) filament. The filaments of high-wattage bulbs are thicker than those of low-wattage bulbs. (It is important to note that both watts and volts are printed on a light bulb. A bulb that is labeled "100W/120V" is 100W *only* if there are 120 volts across it. If there are only 110 volts across it, then the power output is only 84 watts!)

7. A 60-watt bulb and a 100-watt bulb are connected in series in a circuit. Which bulb has

the greater current flowing through it? What if the bulbs are connected in parallel?

Answer: *Note: This is a difficult question.* In a series circuit, each part of the circuit passes the same current. The amount of current depends on the sum of the resistances, and it is the same through each bulb. When connected in parallel, the voltage across each bulb is the same, and the current differs only if the lamp resistances differ. Since high-wattage bulbs have thicker filaments and smaller resistances, the 100-watt bulb will draw more current, whether it is ac or dc.

8. If a 60-watt bulb and a 100-watt bulb are connected in series, which bulb will have the greater voltage drop? What happens when the connection is in parallel?

Answer: *Note: This is a difficult question.* The 100-watt bulb has the thicker filament and lower resistance, so in a series circuit, where the current is the same in each bulb, less energy goes into heat for the lower resistance. This corresponds to lower voltage across the resistance, thus a lower voltage drop. Consequently, the greater voltage drop is across the 60-watt bulb in series. Interestingly enough, in a series connection the 60-watt bulb is brighter than the 100-watt bulb is! When connected in parallel, the voltage across each bulb is the same and the current is greater in the lower-resistance 100-watt bulb, which glows brighter than the 60-watt bulb.

Computational Problems

1. How much current will a person experience if the resistance of their body across 120 volts is 100 000 ohms? If the resistance is lowered to 1000 ohms?

Answers: $I = V/R = 120V/100\ 000$ ohms $= 0.0012$ A, or 1.2 mA. For 1000 ohms, $I = 120V/1000$ ohms $= 0.12$ A, or 120 mA.

2. The resistance of a certain wire is 10 ohms. What would be the resistance of the same wire if it were twice as long? What would be its resistance if it were twice as thick?

Answers: 20 ohms: 2.5 ohms: Twice the diameter has four times the cross-sectional area and one-fourth the resistance.

3. A current of 4 A flows when a resistor is connected across a 12-volt battery. What is the resistance of this resistor?

Answer: $R = V/I = 12V/4A = 3$ ohms.

4. What is the resistance of a clothes iron that draws 8 amperes when connected to 120 volts?

Answer: $R = V/I = 120V/8A = 15$ ohms.

5. How much energy is expended in lighting a 100-Watt bulb for 30 minutes?

Answer: 50 watt-hours, or 0.05 Kwh.

Electric Circuits

To use this planning guide work from left to right and top to bottom.

Chapter 35 Planning Guide		
• The bulleted items are key: Be sure to do them!		

Topic	Exploration	Concept Development	Application
Battery and Bulb		• Text 35.1/Lecture • Demo 35-1	
Electric Circuits	• Act 90 (1 period) Act 92 (1 period)	• Text 35.2/Lecture	
Series Circuits		• Text 35.3/Lecture Demo 35-2 • Con Dev Pract Pg 35-1	Nx-Time Qs 35-1, 35-2
Parallel Circuits		• Text 35.4/Lecture Demo 35-3 • Con Dev Pract Pg 35-2	
Schematic Diagrams		• Text 35.5/Lecture	
Compound Circuits		• Text 35.6/Lecture Con Dev Pract Pg 35-3 Exp 91 (1 period)	
Parallel Circuits and Overloading		• Text 35.7/Lecture	
Video: *Electric Current* **Evaluation:** Chapter 35 Test			

See the Chapter Notes for alternative ways to use all these resources

Objectives

After studying Chapter 35, students will be able to:

• Given a diagram of a battery and a bulb connected by wire, determine whether current will pass through the bulb.

• Distinguish between series circuits and parallel circuits.

• Predict what will happen in a series circuit if there is a break in the wire.

• Relate the current at any point in a series circuit to the current at any other point in the circuit.

• Predict what will happen to the current at any point in a series circuit if an additional device is connected to the circuit.

• Predict what will happen in a parallel circuit if there is a break in any branch of the circuit.

• Relate the current in the lead to a parallel circuit to the current in each branch of the circuit.

• Predict what will happen to the current at any point in a parallel circuit if an additional device is connected to the circuit.

• Interpret a simple schematic diagram of a circuit.

• Given a circuit with two or more devices of equal resistance with some devices connected in series and some connected in parallel, determine the equivalent single resistance for the circuit.

- Explain the cause of overloading household circuits and how to prevent this from happening.

Possible Misconceptions to Correct

- In a parallel circuit, the equivalent resistance of the circuit increases with the addition of more resistors.
- Voltage, rather than only current, flows through a circuit.

Demonstration Equipment

- [35-1] Batteries, bulbs, and connecting wires for all students
- [35-2] A 12-volt automobile battery with brass or copper rods extended from the terminals with alligator clips used to fasten lamps between them. (See the sketch on this page and the comic strip "Parallel Circuit" on page 542 in the text)

Introduction

This chapter is an extension of the previous chapter. It discusses simple series and parallel dc circuits. Complications such as the internal resistance of voltage sources and multiple sources of voltage in series and parallel are avoided. Only simple circuits are treated in this chapter. This way, many students will have a better understanding of simple circuits. More information may benefit a few students, but it will leave too many others with a counter-productive information overload. Since circuits are not the background for any other chapter in the text, this chapter may be omitted from your course. The material is suitable for home study if you do choose to omit it. Students can work in pairs and check each other.

This chapter is best discussed when students have circuit elements in hand. So the suggested lecture is different from previous ones in that it merely guides student activity.

Here's a useful rule for selecting resistors that will always produce an integer when two are combined in parallel. It generates an infinite series of combinations. Values for resistors R_1 and R_2 are given by the formula, $R_1 = xy$, $R_2 = (x-1)R_1$, where x is any integer greater than 1, and y is any integer.

Be sure to make up a battery with the extended terminals demonstration, as shown in the comic strip, *Parallel Circuit*, on text page 542. One way to do this is to fasten binding posts to the terminals of the automobile storage battery and set into them a pair of rigid vertical brass or copper rods. It is easy to see that the 12-volt potential difference between the terminals also occurs across the rods. Next use the alligator clips to attach lamps of equal resistance. Show how the brightness of each lamp, and hence the current through each lamp, does not depend on the number of lamps connected in the parallel circuit. An ammeter placed between one of the rods and a terminal shows line current, which will increase as lamps are added. This is a simple and visually comprehensible demonstration of parallel circuits. You can also show the features of a series circuit by stringing the lamps together and attaching the ends to the extended terminals.

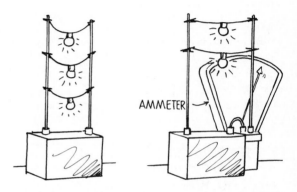

What does it mean when we say that a battery is *dead*? A battery, like anything else, has resistance. If the electrodes in a battery become corroded, the internal resistance prevents the flow of charge, thus we say the battery is dead. Interestingly enough, the dead battery may register full voltage on a voltmeter. This happens because the voltmeter draws only a tiny current, which the battery can easily supply. Drawing a greater current is another story. This situation is similar to what happens in a rusty water pipe. As long as water does not flow in the pipe, a check at a faucet will show full water pressure. However, when the faucet is opened and water flows in the pipe, pressure is reduced by the rusty obstruction. In both cases, the resistance only reduces the pressure or voltage when the current flows. Putting a load on the battery produces an appreciable voltage drop, because too much energy is required to force a charge though its internal resistance.

If the same amount of current flows in a wire and the lamp filament connected to it, why does only the lamp filament glow? The lamp filament glows because of the energy delivered to it. Since the filament has a high resistance, there is a voltage drop across it, and appreciable energy per charge is given to the filament. This energy heats the filament to incandescence. Although just as many electrons flow in the connecting wire, the low resistance of the wire means the electrons carry their energy through the wire, rather than delivering their energy to it. That's what it means to say there is no voltage drop along the connecting wires; energy is not dropped or deposited to them.

Suggested Lecture

Bring out the batteries and bulbs and go to it! Play around with successful and unsuccessful ways to light a bulb (text Figure 35-2).

> DEMONSTRATION [35-1]: This demonstration is not done by you, but rather it is done by your students tinkering around with different ways to light a bulb.

Series Circuit: Make a simple series circuit, as shown in text Figure 35-4, and apply to it the 5 characteristics listed on text page 540. Demonstrate that the circuit is broken when any bulb is loosened.

> DEMONSTRATION [35-2]: Show a series connection between the extended battery terminals.

Parallel Circuit: Make a simple parallel circuit, as shown in text Figure 35-5, and apply to it the 4 characteristics that start at the bottom of page 541.

> DEMONSTRATION [35-3]: Show a parallel connection between the extended battery terminals. Show that the circuit remains in operation when any bulb is loosened. If you attach an ammeter between the battery terminal and the extension rod, you can show how the circuit draws more current as more lamps are added.

Circuit Diagrams: Have your students make up simple circuits with lamps, wire, and a battery. Then have them represent their own circuits with circuit diagrams. The Practice Page for this chapter will be helpful.

Combining Resistors: Note that the numerical values of resistors in the text lend themselves to simple computations. This is done because the important goal is teaching concepts, not handling calculations. Give some advance thought to the resistance values you select for sample circuits. Also, avoid complicated arithmetic that will detract from the concepts. Many students find a pocket calculator helpful.

Home Lighting Circuits: On the board draw a simple parallel circuit using lamps and appliances, such as in text Figure 35-12. Estimate the current flowing through each device, and point out that the current in any branch is not affected when other devices are turned on. Show on your diagram the current in the branches and in the lead wires. Show where the fuse goes and describe its function. Then short your circuit and blow the fuse.

Discuss the consequences of too many appliances operating on the same line. Explain why different sets of lines are directed to various parts of a house. Most house wiring is rated at 30 amperes maximum. A common air conditioner uses about 2400 watts, so if it is operating on 120 volts, the current would be 20 amps. To start operation, more current is needed. (To explain why the starting current is greater would be premature here. If it comes up, you can explain that every motor is also a generator and the input current is met with a generated output that reduces the net current.) If other devices are drawing current on the same line, the fuse will blow when the air conditioner is turned on, so a 220-volt line is usually used for such heavy appliances. Point out that most of the world operates normally at 220-240 volts.

More Think-and-Explain Questions

1. Which will draw more current, a lamp with a thin filament or a lamp with a thick filament? (Assuming the same filament length.)

   Answer: The thick filament has less resistance and will draw (carry) more current than will a thin wire connected across the same potential difference. (It is common to say that a certain resistor "draws" a certain amount of current, but this may be misleading. A resistor doesn't "attract" or "draw" current, just as a pipe in a plumbing circuit doesn't "draw" water. The resistor allows, or provides for, the passage of current when an electrical pressure is established across it.)

2. Does electric current flow *out* of a battery or *through* a battery? Does it flow *into* a light bulb or *through* a light bulb? Explain.

   Answer: Current flows through electrical devices, just as water flows through a plumbing system of pipes. If a water pump produces water pressure, water flows through both the pump and the system. The same is true with electric current in an electric circuit. For example, in a simple circuit consisting of a battery and a lamp, the electric current that flows in the lamp is the same electric current that flows through the wires that connect the lamp. It is also the same electric current that flows through the battery. Current flows *through* these devices. As a side point, it is common to speak of electric current flowing in a circuit, but strictly speaking, it is electric charge that flows in an electric circuit; the flow of charge is current.

3. Which type of current would you expect to power the lamps in your home, ac or dc? What type powers the lamps in an automobile?

Answer: Electric power in your home is probably supplied at 60 hertz via 110-volt to 120 volt electrical outlets. This is ac. It is delivered to your home from transformers between the power source and your home. We will see in Chapter 37 that transformers require ac power for operation. Electric power in a car must be supplied by the battery. Since the positive and negative terminals of the battery do not alternate, the current they produce does not alternate either. Hence current in an automobile flows in one direction and is dc.

Computational Problems

1. How much does it cost to operate a 100-watt lamp continuously for 1 week if the power utility rate is 10 cents per kilowatt-hour?

Answer: $1.68: First, 100 watts = 0.1 kilowatt; secondly, there are 168 hours in one week (7 days x 24 hours/day = 168 hours), so 168 hours × 0.1 kilowatt = 16.8 kilowatt-hours, which at 10 cents per kwh costs $1.68.

2. A 4-watt night light is plugged into a 120-volt circuit and operates continuously for one month. Find the following: (a) the current drawn, (b) the resistance of the filament, (c) the energy consumed in one month, and (d) the cost of operation for one month at the utility rate of 10 cents per kilowatt-hour.

Answers: (a) From power = current × voltage, current = power/voltage = (4W)/(120V) = 1/30 amp. (b) From current = voltage/resistance (Ohm's law), resistance = voltage/current = (120 V)/(1/30 A) = 3600 ohms. (c) First, 4 watts = 0.004 kilowatt; second, there are 720 hours in a month (24 hours/day × 30 days = 720 hours), so 720 hours x 0.004 kilowatt = 2.88 kwh. (d) At the rate of 10 cents per kilowatt-hour, the monthly cost is 2.88 kwh × $.10/kilowatt-hour = $.29.

36 Magnetism

To use this planning guide work from left to right and top to bottom.

Chapter 36 Planning Guide

• The bulleted items are key: Be sure to do them!

Topic	Exploration	Concept Development	Application
Magnetic Poles		• Text 36.1/Lecture • Demo 36-1	Nx-Time Q 36-1 Nx-Time Q 36-2
Magnetic Fields	Act 93 (1 period)	• Text 36.2, 36.3/Lecture	
Magnetic Domains		• Text 36.4/Lecture	
Currents and Magnetic Fields		• Text 36.5/Lecture Demos 36-2, 36-3	
Forces on Moving Charges	• Act 94 (1 period)	• Text 36.6/Lecture • Con Dev Pract Pg 36-1 Demo 36-4	
Forces on Wires		• Text 36.7/Lecture Demo 36-5	Nx-Time Q 36-3
Meters and Motors		• Text 36.8/Lecture Demos 36-6, 36-7	
Earth's Magnetic Field		• Text 36.9/Lecture	

Video: *Magnetism*
Evaluation: Chapter 36 Test

See the Chapter Notes for alternative ways to use all these resources

Objectives

After studying Chapter 36, students will be able to:

• Describe the differences and similarities between magnetic poles and electric charges.

• Interpret the strength of a magnetic field at different points near a magnet by using the pattern formed by iron filings.

• Relate the motion of electrons within a material to the ability of the material to become a magnet.

• Describe what happens to the magnetic domains of iron in the presence of a strong magnet.

• Explain why magnets lose their magnetism when dropped or heated.

• Describe the magnetic field produced by a current-carrying wire and give examples of how the field can be made stronger.

• Describe the conditions necessary for a magnetic field to exert a force on a charged particle in the field.

• Describe some practical applications of a magnetic field exerting a force on a current-carrying wire.

• Suggest possible causes for the earth's magnetic field.

Possible Misconceptions to Correct

- Magnetic poles are to magnets what electric charge is to electricity.
- Magnetic poles move in iron the same way electrons move in electrical conductors.
- The magnetic force on charged particles, like the electric force on charged particles, is in the direction of the magnetic field, rather than at right angles to both the field and the velocity of the charge.
- Like an electric field, a magnetic field can increase the speed of charged particles.

Demonstration Equipment

- [36-1] Overhead projector, iron filings, several magnets, and a sheet of transparent plastic
- [36-2] Compass and wire that carries dc current
- [36-3] Items for constructing an electromagnet: iron bar, wire, and a battery
- [36-4] Oscilloscope or television monitor and a magnet to distort image
- [36-5] DC current-carrying wire and horseshoe-shaped magnet
- [36-6] Large meter (galvanometer, ammeter, voltmeter) with a visible coil to show rotation in the magnetic field
- [36-7] DC demonstration motor with power source

Introduction

This chapter, like so many others, links the subject matter to the environment. The material in this chapter is a prerequisite for the next chapter.

Note that the text avoids the confusion of the north geographic pole of the earth being a south magnetic pole and the south geographic pole being a north magnetic pole.

Microscopic views of magnetic domains are shown in the Ealing film loop #A80-2033/1, *Ferromagnetic Domain Wall Motion*.

Details of a do-it-yourself paper clip electric motor are available. If you want a copy, I will be happy to send you one. (Address: Paul Hewitt, City College of San Francisco, SF, CA 94112)

Suggested Lecture

Begin by holding a magnet above some nails or paper clips on your lecture table. State that the nails or paper clips are flat on the table because every particle of matter in the planet Earth is gravitationally pulling the nails and clips to the table. Then show that your magnet will outpull the entire earth and lift the nails or clips off the table.

Magnetic Field: Introduce the concept of magnetic field with the following demonstration.

> DEMONSTRATION [36-1]: Show the field patterns around bar magnets with the use of an overhead projector and iron filings. Simply place a magnet on the glass surface of the projector and cover it with a sheet of plastic. Then sprinkle iron filings on the plastic. (This is shown for a bar magnet in text Figure 36-4 and for pairs of magnets in Figure 36-6.)

CHECK QUESTION: How do the field lines of magnetic, electric, and gravity fields differ? [Magnetic field lines form closed loops. Conventionally, they circle from the north pole to the south pole outside a magnet and from the south pole to the north pole inside the magnet. Electric field lines emanate from positive charges toward negative charges. Gravity field lines emanate only from mass.]

Magnetic Domains: Discuss the source of magnetism — the motion of charges. All magnetism starts with a moving electric charge; in the spin of the electron about its own axis, in the revolution of the electron about the nuclear axis, and as the electron drifts as part of an electric current. It should be enough to simply acknowledge that the magnetic field is a relativistic "side effect" or "distortion" in the electric field of a moving charge. (Unless you've already treated special relativity in great detail, the relativistic explanation may be too involved to be effective.)

Describe magnetic induction and show how moving an unmagnetized nail near a magnet induces it to become magnetized and attracted to the other materials. Next contrast magnetizing iron to magnetizing a piece of aluminum. Discuss unpaired electron spins and magnetic domains. Compare text Figures 32-12 and 32-13, showing bits of paper attracted to a charged object and a charged balloon stuck to a wall by electrostatic induction, to the magnet and nails in text Figure 36-10. Point out the similarities of inducing charge polarization electrically and inducing the alignment of magnetic domains magnetically.

CHECK QUESTION: Given that charges are either positive or negative, are magnets either north or south? [No, every magnet has both a north pole and south pole just as every coin has two sides. Every pole is either north or south and every individual charge is either a positive or a negative. However, a single magnet is a pole pair, namely north and south. Be sure that your students don't confuse the positive and negative charges of electricity with magnetism. (Some theoretical physicists believe there are "monopoles," elemental magnetic particles, north or south. At this writing, no evidence supports this theory.)]

Electric Currents and Magnetic Fields:
Show the interaction of a current-carrying wire and a magnet.

DEMONSTRATION [36-1]: Place a wire without current near a compass needle, as text Figure 36-11 suggests. Show that current in the wire deflects the compass needle. (This was Orested's classroom discovery.)

Here's an interesting sidelight. When the magnetic field around a current-carrying conductor is undesirable, double wires are used with the return wire right next to the wire. The net current for the double wire is zero and no magnetic field surrounds it. Wires are often braided to combat slight magnetic fields where the cancellation nearby is not perfect.

Explain how the magnetic field lines crowd up in a loop (text Figure 36-12), and then multiple loops (text Figure 36-13 right). Now you have an electromagnet.

DEMONSTRATION [36-2]: Make a simple electromagnet in front of your class by winding wire around a nail and picking up paper clips with it. Compare this to the junk-yard magnet shown in text Figure 36-1, then show other types of magnets. Discuss solenoids.

Acknowledge that magnets can repel as well as attract and discuss the application of magnetic repulsion to high-speed passenger trains (text Figure 36-14).

Magnetic Forces on Moving Charges:
Perform the following demonstrations in the order indicated to emphasize that a force acts on electric charges that are moving through a magnetic field.

DEMONSTRATION [36-3]: Show how a magnet distorts the beam of an oscilloscope or TV picture. Stress the role of motion.

Discuss the motion of a charged particle injected into a magnetic field perpendicular to the field lines and explain how the beam moves in a circular path. The perpendicular push is a centripetal force that acts along the radius of its path. Briefly discuss cyclotrons and bevatrons that have radii ranging from less than a meter to more than a kilometer. Note that since the magnetic force on moving charges is always perpendicular to velocity, there is never a component of force in the direction of motion. This means a magnetic field cannot do work on a moving charge. A magnetic field can only change the direction of the charge. Consequently, in cyclotrons and bevatrons, electric fields accelerate the charges and increase their kinetic energies; the magnetic fields simply guide the path of the charge.

Forces on Current-Carrying Wires:
It is a simple step from the deflection of charges to the deflection of the wires that enclose these deflected charges.

DEMONSTRATION [36-4]: Show how a wire jumps out of (or into) a magnet when current is passed through it (text Figure 36-17). Reverse the current or turn the wire around to show both cases.

Electric Meters and Electric Motors:
Extrapolate the previous demonstation to the same deflection occurring in galvanometers, ammeters, and voltmeters.

DEMONSTRATION [36-5]: With your largest meter (galvanometer, ammeter, voltmeter), point out to your class the coil of wire that is suspended in the magnetic field of the permanent magnet (see text Figure 36-19).

Now you are ready to extend this idea to the electric motor.

DEMONSTRATION [36-6]: Show the operation of a dc demonstration motor.

Earth's Magnetic Field:
Discuss the field pattern about the earth and how cosmic rays are deflected by the magnetic field. In discussing pole reversals, add that the magnetic field of the sun undergoes reversals about every eleven years.

Not mentioned in the text are the Van Allen radiation belts. These belts consist of two lunar-shaped shells. They are composed of charged particles trapped in the earth's magnetic field a few hundred to at least 50 000 kilometers above the earth's surface. Protons and electrons are found in both the outer part and the inner part of the shells. The inner portion is now partially masked by newer electrons that were produced by high altitude nuclear explosions in 1962. In a ring around the poles, ions and electrons dip into the atmosphere and cause it to glow like a fluorescent lamp. This is the beautiful Aurora Borealis seen in the Northern Hemisphere and the Australis Borealis seen in the Southern Hemisphere.

The magnetic field of the earth protects us from much of the cosmic rays we would receive otherwise. Cosmic ray bombardment is at a maximum at the poles because incoming particles do not criss-cross the earth's field (they would be deflected), but rather they follow the field lines and are not deflected. At sea level at the equator, incidence of cosmic rays averages from one to three particles per square centimeter per minute; this number increases rapidly with altitude. This is a main reason for the relatively short working hours of flight personnel in high-flying aircraft. Two cross-country round trips expose us to a radiation dosage equivalent to a chest X ray (more about this in the Chapter 39 material in this *Teaching Guide*).

The probable role of the earth's magnetic field in evolution is not treated in the chapter material, but it is considered in Think and Explain Question 11 in the text. One theory is that when life was passing through its earliest phases, the magnetic field of the earth was strong enough to hold off cosmic and solar radiations that were violent enough to destroy life. During pole reversals, cosmic radiation on the earth's surface was increased, aided by the spilling of the particles in the Van Allen belts. This increased the incidence of mutation of the primitive life forms that existed. Sudden bursts of radiation may have been as effective in changing life forms as X rays have been in the famous heredity studies of fruit flies. The coincidences of the dates of increased life changes and the dates of the magnetic pole reversals lend support to this theory.

Not in the chapter material, but referred to in Think and Explain Question 10 in the text, are the multiple-domain magnetite magnets within the skulls of pigeons. These are connected by a large number of nerves to the pigeon brain. For several years it has been known that single-domain magnetite grains strung together to form internal compasses exist in certain bacteria. Bacteria south of the equator build the same single-domain magnets as their counterparts north of the equator do, but the magnetite grains are aligned in the opposite direction to coincide with the oppositely-directed magnetic field in the Southern Hemisphere. Magnetic material has also been found in the abdomen of bees. To date human beings show no such magnetic materials.

Not in the text is MRI — magnetic resonance imaging — formerly called NMR, nuclear magnetic resonance. This is relatively new and widely-used in medicine, particularly as a method of cancer detection. An external alternating magnetic field is applied to a part of the body of a patient. Due to the environment of neighboring atoms, slight differences in the natural frequencies of magnetic quadrupole moments of atomic nuclei, commonly protons, are detected by a "magnetic echo." The resonant signals from the nuclei of atoms in normal cells differ slightly for cancerous tissue and are picked up by a sensitive magnetometer.

More Think-and-Explain Questions

1. The core of the earth is probably composed of iron and nickel, which are excellent metals for making permanent magnets. Why is it unlikely that the earth's core is a permanent magnet?

   Answer: The iron and nickel that compose the earth's core is too hot for permanent alignment of magnetic domains, and therefore the core does not become a permanent magnet. The earth's magnetism probably originates in the electric currents surrounding the earth's core.

2. One way to make a compass is to place a magnetized needle into a piece of cork and float it in a plastic bucket of water. The needle will align itself with the earth's magnetic field. Since the north pole of this compass is attracted northward, will the needle float toward the northward side of the bucket? Defend your answer.

   Answer: The needle is not pulled toward the north side of the bucket, because the south pole of the magnet is equally attracted southward. The net force on the needle is zero.

3. What is the net magnetic force on a compass needle? By what mechanism does a compass needle line up with a magnetic field?

   Answer: The net force on a compass needle is zero, because its north and south poles are pulled in equal and opposite directions in the earth's magnetic field. When the needle is not aligned with the magnetic field of the earth, a pair of torques is produced. (This is because each end of the needle is pulled in an opposite direction, but the lines of action of these forces do not pass through the needle's axis of rotation.) This pair of equal torques, called a *couple*, rotates the needle into alignment with the earth's magnetic field.

4. Can an electron be set into motion with a magnetic field? An electric field?

   Answer: Relative motion between an electron and a magnetic field is required for a magnetic force on an electron to occur; that is, either the electron moves through a stationary magnetic field or a magnetic field moves by an electron, or both. An electron at rest between the poles of a magnet will experience no magnetic force. In an electric field, however, an electron will be accelerated whether or not motion is involved.

To use this planning guide work from left to right and top to bottom.

Chapter 37 Planning Guide
• The bulleted items are key: Be sure to do them!

Topic	Exploration	Concept Development	Application
Electromagnetic Induction		• Text 37.1/Lecture • Demo 37-1	Nx-Time Q 37-2
Faraday's Law	Act 95 (1 period)	• Text 37.2, 37.3/Lecture Con Dev Pract Pg 37-1	
Generators and AC		• Text 37.4/Lecture	
Motor and Generator Comparison		• Text 37.5/Lecture Demos 37-1, 37-2	
Transformers		• Text 37.6/Lecture • Con Dev Pract Pg 37-2 Demo 37-3	Nx-Time Q 37-1
Power Transmission		• Text 37.7/Lecture Demo 37-4	• Con Dev Pract Pgs 37-3, 37-4
Induction of E&M Fields		• Text 37.8/Lecture Demos 37-5, 37-6	
E&M Waves		• Text 37.9/Lecture	

Video: *Electromagnetic Induction*
Evaluation: Chapter 37 Test

See the Chapter Notes for alternative ways to use all these resources

Objectives

After studying Chapter 37, students will be able to:

• Describe how voltage is induced in a coil of wire.

• Relate the induced voltage in a coil to the number of loops in the coil and the rate of change of external magnetic field intensity through the loops.

• Describe a generator and explain how it works.

• Compare and contrast the motor effect and generator effect.

• Describe a transformer and explain how it works.

• Explain why transformers are used for transmission of electric power.

• Relate the magnitude and direction of an induced electric field to the inducing magnetic field.

• Relate the magnitude and direction of an induced magnetic field to the inducing electric field.

• Explain how the electric and magnetic fields of an electromagnetic wave regenerate each other so that the wave pattern moves outward.

Possible Misconceptions to Correct

- Voltage is produced by a magnet, rather than by the work done when a magnet and closed loop(s) of wire are moved relative to each other.
- A generator and a motor are fundamentally different from each other.
- A transformer can step up energy, or step up power.

Demonstration Equipment

- [37-1] Galvanometer, loop of wire, horseshoe magnet

- [37-2] Same as above with wire that can be bent into a coil or attached to prewound coils
- [37-3] Demonstration motor that can be shown as a generator, with galvanometer (Genecon will do)
- [37-4] Hand-cranked generator and a lamp
- [37-5] Step-down transformer with nail to weld when connected to an ac source
- [37-6] Elihu Thompson electromagnetic apparatus shown in the sketch on page 175 with lamp on waterproof coil that you immerse in a glass of water
- [37-7] Elihu Thompson electromagnetic apparatus and aluminum ring that levitates or jumps when power is applied

Introduction

This chapter focuses on the important features of electromagnetic induction. The chapter should be supported by various lecture demonstrations of electromagnetic induction, such as those in the figures. Difficult concepts such as reactance, back emf, Lenz's law, and the left-hand and right-hand rules are not covered. The focus here is transmitting energy from one place to another without physical contact.

If you have a Tesla coil, demonstrate induction by lighting up a disconnected fluorescent lamp a meter or more away. This is impressive.

Two computer programs, *Radiating Dipole* and *Moving Charge* from the computer disk *Good Stuff!*, nicely complement the material in this chapter. *Radiating Dipole* shows the emanation of electric field lines from an antenna. *Moving Charge* shows how all kinds of electromagnetic radiation are produced by accelerating charges.

Suggested Lecture

In the previous chapter we discussed the production of electricity from magnetism. During the first half of the 1800s physicists asked "Can it be the other way around? Can we start with magnetism and produce electricity?" Indeed we can — enough electricity to light entire cities! Present a galvanometer, magnet, and wire loop to the students. Do this well away from the electric power sources of your previous lecture.

Faraday's Law: This lecture is a series of demonstrations.

DEMONSTRATION [37-1]: Produce motion between a wire loop and a magnet as

shown in text Figure 37-2, preferably with a large galvanometer meant for classroom demonstration.

Follow this demonstration with one using a bar magnet and coil.

DEMONSTRATION [37-2]: Plunge a bar magnet in a coil as shown in text Figure 37-3. Show that twice as much deflection occurs for a coil with twice the number of the turns, and so on. Establish that a directly-proportional relationship exists between induced voltage and number of turns in the coil.

Summarize the foregoing demonstrations and cite Faraday's law of electromagnetic induction. Emphasize the importance of this discovery by Faraday and Henry and how its application transformed the world. Ask students what it would be like to have no electric lights — to live in a time when illumination after the sun goes down is only by candles and whale-oil lamps. Also, ask them to imagine not having electric motors! Point out that in our older cities, many buildings still have pre-electric light fixtures, such as gas and oil lamps.

Point out that the magnet is not a source of voltage. Voltage is induced when work is done to push the magnet into the coil. It may seem that voltage is obtained by simply increasing the number of loops in a coil. This is done at the expense of making it more difficult to push the magnet into more loops (see text Figure 37-4). It turns out that the current induced is surrounded by its own magnetic field, which resists the magnet being pushed or pulled. Consequently, when greater current is induced by the action, more resistance is met. This is evident when cranking a generator when an additional electrical load is suddenly introduced. For example, driving a car with the lights on uses more fuel.

Discuss the operation of traffic control signals that are activated by metal vehicles passing over wire loops imbedded in the road surface (see Think and Explain Question 9 in the text). Also discuss the metal detectors used in airports. The passage of ferromagnetic material through or over these loops alters the magnetic field and induces voltage in the loops.

Generators:

DEMONSTRATION [37-3]: Return to the motor you previously demonstrated and show that when you apply mechanical energy, it becomes a generator (a Genecon will do).

Compare a motor and a generator, which are the same in principle. When the input is electric energy, it is converted to mechanical energy — a motor. When mechanical energy is put in, it is converted to electrical energy — a generator. In fact, a motor also acts as a generator by creating a "back voltage" (back emf) and an opposing current. The net current in a motor is the input current minus the generated back current. The net current in a power saw will not cause it to overheat and damage the motor windings — as long as the motor is running and generating a back current that keeps the net current low. However, suppose you should jam the saw so that it can't spin. Without the back current generated by the spinning armature, the net current becomes dangerously high and can burn out the motor.

Interestingly enough, electric motors are used in diesel-powered railroad engines. The combustion engine cannot bring a heavy load from rest, but an electric motor can. Why? This occurs because when the armature is not turning, the current in the windings is huge and has a corresponding huge force. As both the train and the motor gain speed, the back current generated by the motor brings the net current in the motor down to non-overheating levels.

DEMONSTRATION [37-4]: Light a bulb with a hand-cranked generator and show that it turns easier when the bulb is loosened and the load is removed. Allow students to try this themselves either during class or after class.

Stress again the fact that we don't get something for nothing with electromagnetic induction, and point out text Figure 37-4. This can be readily felt when by switching on lamps that are powered with a hand-cranked or a bicycle generator. Each student should experience this. The conservation of energy reigns!

In discussing the operation of a generator via text Figures 37-6 and 37-7, point out that maximum voltage is induced not when the loop contains the most magnetic field lines, but rather when the greatest number of field lines are "clipped" (changed) as the loop is turned. Hence in Figure 37-7, the voltage is at its maximum when the loop passes through the zero-number-of-lines point. Its rate of change of magnetic field lines is greatest at this point.

Tell students that with the advent of the generator, the task at hand was to design a way to move coils of wire past magnetic fields, or to move magnetic fields past coils of wire. Turbines were placed beneath waterfalls, and steam from boiling water was used to keep turbine blades turning as electricity was generated in large amounts — enter the industrial revolution.

Transformers: Explain how a transformer works. Some students may be very confused about the seeming contradiction with Ohm's law — the idea that when voltage in the secondary is increased, current in the secondary is decreased. Make clear that when the voltage in the coil of the secondary and the circuit it connects is increased, the current in *that* circuit also increases. The decrease is with respect to the current that powers the *primary*. So $P = IV$ does not contradict Ohm's law!

DEMONSTRATION [37-5]: With a step-down transformer, weld a pair of nails together. This is a spectacular demonstration when you casually place your fingers between the nail ends before they make contact. Then after removing your fingers bring the points together allowing the sparks to fly while the nails quickly become red and then white hot.

Cite the role of the transformer in stepping down voltages in toy electric trains, power calculators, and portable radios. Also cite the role of stepping up voltages in TV sets and various electrical devices, in addition to both stepping up and stepping down voltages in power transmission. Stress that in no way is energy or power stepped up or down — a conservation of energy no no! Carefully go over the comic strip "Power Lines," on text page 577. The physics here is deeper than in the other comic strips and you may need to elaborate.

Induction of Fields: Continue with the switch to ac and bring out the classical Elihu Thompson Electromagnetic Demonstration Apparatus, as shown in the sketch.

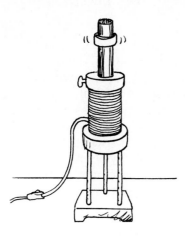

DEMONSTRATION [37-6]: Show the classic lighting of the lamp in a jar of water. Impressive! (The lamp is mounted on a waxed, waterproof coil that intercepts the changing magnetic flux of the device, induces current, and illuminates the lamp. The water serves no purpose other than making the demonstration more interesting.)

DEMONSTRATION [37-7]: With the power on, levitate an aluminum ring over the extended pole of the Elihu Tompson device.

CHECK QUESTION: Do you know enough physics to state how much electromagnetic force supports this 1-Newton aluminum ring (assume the ring weighs 1 N)? [1 N, which comes more from a knowledge about forces in general (based on Newton's laws), rather than from a knowledge of electromagnetic forces. Since the ring is at rest and not accelerating, the upward electromagnetic force (in newtons!) must be equal to the downward force of gravity.]

With the power off, place the ring at the base of the extended pole. When you switch on the power the current induced in the ring via electromagnetic induction converts the ring into an ac electromagnet. (According to Lenz's law, which is not covered in the text, the polarity of the induced magnet always opposes the magnetic field imposed.)

CHECK QUESTION: Do you know enough physics to state whether or not the electromagnetic force that popped the ring was more, equal to, or less than the magnetic force that produced levitation earlier? [It is more, because it accelerated upward, as shown by the upward force, which was more than the weight. This is also understandable because the ring was lower and was intercepting more changing magnetic field lines.]

State that underlying all that was discussed and observed is something more basic than voltages and currents. It is the induction of fields, both electric and magnetic. Because this is true, we can send signals without wires (radio and TV). Furthermore, energy reaches us from the sun in the form of sunlight.

Here's a point not stated in the text: The concept that a change in either field induces the other led Einstein to develop his special theory of relativity. Einstein showed that a magnetic field results when an electric field is seen by a moving observer, and that an electric field results when a magnetic field is seen by a moving observer. The fields are relative.

Power Transmission: Not mentioned in the text is that cost is a main reason for high-voltage power lines. If higher currents were carried in the lines, the wires would have to be thicker and therefore would be costlier. They would also be heavier and would require stronger towers.

Electromagnetic Waves: Ask your class to recall your recent demonstration of charging a rubber rod with cat's fur. When you brought the rod near a charged pith ball, you produced action at a distance. When you moved the charged rod, the charged ball moved also. If you gently oscillated the rod, the ball in turn oscillated. State that one can think of this behavior as either action-at-a-distance or the interaction of the ball and rod with the surrounding space — the electric field. For low frequencies, the ball will swing in rhythm with the shaking rod. However, the inertia of the ball and its pendulum configuration causes a poor response when the rod is shaken. Consequently, it's best not to actually show this, but only to describe it and go through the motions as if the equipment were present. You can easily establish in your students' minds that is reasonable for the ball to move back and forth in response to the shaking (changing) electric field around the shaking rod. Carry this further by considering the ball as simply a point charge with negligible mass. Now it will respond in synchronous rhythm to the shaking rod. Increase the frequency of the shaking rod and state that not only is there a shaking electric field about the rod, but because of its changing, there is now a different kind of field.

CHECK QUESTIONS: What kind of field is induced by the shaking rod? What kind of field in turn, does this induced field induce? And further in turn, what kind of field does this further induced field induce? And so on. [The shaking charge induces a magnetic field, which changing in turn induces an electric field, etc. The result is an electromagnetic wave.]

Cite the idea of the optimum speed of field disturbances from the shaking rod to the ball. This is consistent with energy conservation, see text page 580.

Review the family of electromagnetic waves back in Figure 27-5 on text page 399.

CHECK QUESTION: Why are sound waves not included in Figure 27-5? [Sound is a mechanical wave. It is not electric or magnetic, and it is not part of the electromagnetic spectrum!]

If misconceptions are still evident, pose the following question.

CHECK QUESTION: Give the correct answer. The speed of radio signals from the antenna of the local broadcasting station are (1) usually less than 340 m/s, (2) about 340 m/s, depending on atmospheric conditions, (3) always greater than 340 m/s, or (4) none of the above! [The correct choice is 3, because the speed of an electromagnetic signal is 3×10^8 m/s, which is considerably more than 340 m/s! (In this case, there is no exception to the "always" nature of this question.)]

Tell students that the room they sit in is chock full of waves of many frequencies. You might turn out the lights and state that the total amount of radiation in the room decreased very little as a result. Explain that light waves make up a tiny part of the vibrations that engulf us at every moment.

Next discuss how until the present century, reality was what people could see and touch. As knowledge of the electromagnetic spectrum became common, people learned that what they can see and touch is less than one millionth of reality. As Buckminster Fuller once noted; ninety-nine percent of all that is going to affect our tomorrows is being developed today by humans using instruments that work in ranges of reality that are not humanly sensible.

More Think-and-Explain Questions

1. What is the basic difference between an electric generator and an electric motor?

   Answer: There is no basic difference between an electric motor and electric generator. When mechanical energy is put into the device and electric energy comes out, we call it a generator. When electrical energy is put in and mechanical energy comes out, we call it a motor. (There are, however, many practical differences in designs of motors and generators.)

2. Why is it important that the core of a transformer pass through both coils?

   Answer: High efficiency requires that the maximum number of magnetic field lines produced in the primary are intercepted by the secondary. The core guides the lines from the primary through the secondary. Transformed power would be needlessly lost otherwise.

3. What causes the hum often heard when a transformer is operating?

   Answer: The hum heard when a transformer is operating on a 60 hertz ac line is a 60 hertz forced vibration of the iron slabs in the transformer core as the magnetic polarities alternate. The hum is greater if resonance causes a vibration in other mechanical parts.

4. If a bar magnet is thrown into a coil of wire, it will slow down. Why?

   Answer: The moving magnet will induce a current in the loop. The field so produced tends to repel the magnet as it approaches and to attract it as it leaves, slowing it in its flight. From an energy point of view, the energy of the current induced in the loop is equal to the loss of kinetic energy of the magnet

5. What is the source of all electromagnetic waves?

   Answer: An accelerating electric charge

38   The Atom and the Quantum

To use this planning guide work from left to right and top to bottom.

Chapter 38 Planning Guide
• *The bulleted items are key: Be sure to do them!*

Topic	Exploration	Concept Development	Application
Models		• Text 38.1/Lecture	• Demo 38-1
Light Quanta		• Text 38.2/Lecture	
Photoelectric Effect	Act 96 (1 period)	• Text 38.3/Lecture Demos 38-1, 38-2	Nx-Time Q 38-1
Wave-Particle Duality		• Text 38.4 – 38.6/Lecture • Con Dev Pract Pg 38-1	
Sizes of Atoms		• Text 38.7/Lecture	Nx-Time Q 38-2
Quantum Physics		• Text 38.8/Lecture	

Video: *None*
Evaluation: Chapter 38 Test

See the Chapter Notes for alternative ways to use all these resources

Objectives

After studying Chapter 38, students will be able to:
• Give examples of models for the atom and for light.
• Explain why the energy of light can be considered a multiple of small units of energy.
• Relate the energy of a photon to its frequency.
• Explain why the photoelectric effect is evidence for the particle nature of light.
• Cite evidence for the wave nature of electrons.
• Describe de Broglie's model of matter waves in the atom and use it to explain the lines seen in atomic spectra.
• Explain why the diameter of heavy elements are not much larger than the diameter of lighter elements.
• Describe the limits of Newton's laws of motion.

Possible Misconceptions to Correct

• If something is a wave, it can't be a particle, and vice versa.
• The heavier the atom is, the larger it is in size.
• Quantum physics applies only to the microscopic world.

Demonstration Equipment

• [38-1] A box about the size of a cigar box, a tennis ball, a plastic foam ball about the same size that is painted red, and a massive ball painted blue
• [38-2] Photoelectric effect apparatus: electroscope, freshly polished piece of zinc, open carbon-arc lamp, quartz lens to focus light on the zinc, cat's fur, and silk

Introduction

Although this chapter begins Unit VI, it is a continuation of Chapters 27 and 28 to some extent. It could well follow the last part of Chapter 28, Section 28.11, *The Atomic Color Code — Atomic Spectra*. This chapter can, with some discussion of atomic spectra, stand on its own as a continuation and conclusion of Chapter 17, *The Atomic Nature of Matter*. For a short course, Chapter 9 followed by this chapter should work quite well. It is background for Chapters 39 and 40, but is not a prerequisite.

For more information about the physicists who took part in the development of quantum mechanics, read Barbara Cline's book, *The Questioners: Physicists and the Quantum Theory*, (Crowell, NY, 1973). For more on modern physics, I recommend *The Cosmic Code: Quantum Physics as the Language of Nature*, by Heinz Pagels (Simon & Schuster, NY, 1982). *Sympathetic Vibrations: Reflections of Physicists as a Way of Life* by K.C. Cole (Morrow, NY, 1984) delightfully unravels some of the theories of Bohr, Einstein, and other developers of quantum physics.

The computer program *Free Particle* from the computer disk *Good Stuff!* simulates the appearance of a small particle moving in terms of wave motion and nicely complements this chapter.

Suggested Lecture

Recall the treatment of atomic spectra in Chapter 27. We learned that different colors of light have different frequencies (red, lowest and violet, highest). Below red is the infrared (IR) part of the electromagnetic spectrum and above violet, the ultraviolet (UV). State that these different frequencies have different energies (red, lowest and violet, highest). Point out that the high-energy UV components of sunlight are responsible for sunburn. Summarize this on the board by writing the proportion, $E \sim f$. The energy of a packet of light energy, a *photon*, is directly proportional to its frequency.

> CHECK QUESTION: In a photographer's darkroom, low-energy photons that won't expose film are required. What color light is best for a safety light in a darkroom? [Red]

State that the ratio of the energy E of light to its frequency f is always the same, 6.67×10^{-34} joule seconds, abbreviated h, Planck's constant. The proportion is expressed as an exact equation, $E = hf$. (You may compare this with Hooke's law, $F \sim x$, where k is the proportionality constant that makes $F = kx$ an exact equation.)

Photoelectric Effect: Einstein is best remembered for his theories of relativity, but his Nobel Prize in Physics was for the photoelectric effect.

> DEMONSTRATION [38-1]: Show a simulated photoelectric effect with a tennis ball in a small shallow box about the size of a cigar box. The tennis ball represents an electron. Toss a red plastic foam ball at the tennis ball to show that the foam ball does not have enough energy to knock the tennis ball out of the box. Then toss a more massive blue ball into the box and the tennis ball pops out The red foam ball represents a photon of red light (a small KE). The blue ball represents a photon of blue or violet light (a greater KE). Explain that high-energy photons knock electrons from the material, whereas low-energy photons don't. You can simulate the work function of the material by adjusting the depth of the box with slabs of material that cover the bottom. It is more difficult to dislodge a deeply-set ball than to dislodge one resting on a slab that is nearly as thick as the box is deep.

> DEMONSTRATION [38-2]: Show the actual photoelectric effect by placing a freshly polished piece of zinc on a charged electroscope and illuminate it with an open carbon arc lamp (no glass lens). To focus the beam, use a quartz lens to pass the UV light. Show that a positively-charged electroscope will not lose its charge when the light shines on the zinc plate, but a negatively-charged electroscope will discharge quickly in the same light. Electrons are ejected from the zinc surface. Block the UV light with a glass plate and the discharge ceases. If you have a quartz prism, go further and pass the light through a slit, then through the prism onto the zinc. Show that the negatively-charged electroscope discharges only when the portion of the spectrum beyond the violet end strikes the zinc plate.

Wave-Particles: Often there is a lot of confusion about the wave-particle duality. Light behaves like a wave when it travels in empty space, and it behaves like a particle when it interacts with solid matter. It is incorrect to insist that light must be both a particle and a wave at the same time. What something *is* and what it *does* are not the same.

Planck's constant surfaces again in the de Broglie formula, which relates the wavelength of a "matter wave" to its momentum. Like light, matter traveling through space has wave properties. When incident upon a target, the particle nature of light is evident. We don't ordinarily notice the wave nature of matter simply because the wavelength is so extremely small. The footnote on text page 588 illustrates this.

The matter-wave concept gives a clearer picture of the electrons that "circle" the atomic nucleus. Instead of picturing them as tiny BBs whirling like planets, the matter-wave concept suggests we see the electrons as a smear of standing waves of energy. The smeared electron matter-waves exist only in orbits that allow constructive interference (see text Figures 38-8 and 38-9). Explain that if the distance around the orbit (the circumference) equals an integral number of wavelengths, constructive interference of the electron matter-waves is possible. This explains why the orbits have the discrete radii described by Bohr. The paper-clip analogy on page 590 of the text illustrates this concept.

Atomic Sizes: Draw a model of the hydrogen atom on the board. Then place a second positive charge in the nucleus and ask what will happen to the force that holds the electron in orbit. [The force on the electron will double.] What will this doubled force do to the size of the orbit (wave or particle)? [It will pull the electron in tighter.] The double charge can also hold an extra electron and form helium. It is understandable that helium is a smaller atom than hydrogen. (No wonder it leaks so readily through balloons!) Similarly, uranium's innermost orbits (wave model or particle model) are close to the nucleus, and the uranium atom is only about three times the diameter of the hydrogen atom.

You might also mention that elements shrink in size as one goes from left to right in the periodic table. Fluorine, for example, is considerably smaller than the atoms preceding it in its row. Because of their small size, fluorine atoms can penetrate the enamel of teeth. Once inside, fluorine adds strength to the teeth like steel girders strengthening a building.

Quantum Physics: Distinguish between *classical physics* and *quantum physics*. Point out that classical physics is primarily physics before 1900 that focused on the study of familiar things, such as the forces, motions, momenta, and energy of massive particles that behave in a predictable manner in accordance with Newton's laws. For this reason, classical physics is often called Newtonian physics. After 1900, physicists found that Newtonian rules simply don't apply in the domain of the very small (the submicroscopic). This is the realm of quantum physics, where everything is "grainy" and where values of energy, momentum, position, and perhaps even time occur in lumps, or quanta, and where all are governed by probabilities rather than certainties.

Clarify that *quantum*, (plural, *quanta*) means a quantity that occurs or changes only in elemental "lumps." The total weight of a bag of pennies is quantized in the sense that its weight is a whole multiple of the weight of a single penny. Electric charge is quantized, for the charge on any body is a whole number multiple of the charge of a single electron (quarks notwithstanding). A chunk of gold is quantized in that it is made of a whole number of gold atoms. This is old stuff. The new stuff is that energy, angular momentum, and perhaps even time, are composed of "lumps." It seems that everything is "grainy" when carefully scrutinized. Everything is made of "quanta."

Quantum physics is concerned with the extremely small. Today's physicists, after all, are involved in exploring extremes, such as the outer limits of fast and slow, hot and cold, few and many, and big and small.

The philosophical implications of quantum physics is left to your lecture. You may or may not have much to say about this. You should point out that quantum physics is not as well defined as other bodies of knowledge that are less complex or more easily studied. It has holes and is still regarded by many physicists as an incomplete theory.

More Think-and-Explain Questions

1. Distinguish between classical physics and quantum physics.

   Answer: Classical physics is primarily physics before 1900 that involves the study of familiar things such as the forces, motions, momenta, and energy of massive particles that behave in a predictable manner in accordance with Newton's laws. Classical physics is often called Newtonian physics.

2. We don't notice the wavelength of moving matter in our ordinary experience. Is this because the wavelength is extraordinarily large or extraordinarily small?

   Answer: The momenta of moving things in our everyday environment are huge compared to the momenta of submicroscopic particles, even at speeds near the speed of light. This is because the masses are so huge in comparison. The large momenta, in accordance with de Broglie's formula, correspond to incredibly short wavelengths. See the footnote on text page 588.

3. A friend says, "If an electron is not a particle, then it must be a wave." What is your response?

   Answer: We don't know if an electron *is* a particle or a wave. We know that it *behaves* like a wave when it moves from one place to another, and it behaves like a particle when it is incident upon a detector. The unwarranted assumption is that an electron must *be* either a particle *or* a wave.

39 The Atomic Nucleus and Radioactivity

To use this planning guide work from left to right and top to bottom.

Chapter 39 Planning Guide
• *The bulleted items are key: Be sure to do them!*

Topic	Exploration	Concept Development	Application
Atomic Nucleus		• Text 39.1/Lecture	• Demo 39-1
Radioactive Decay		• Text 39.2/Lecture	Nx-Time Q 39-1
Penetrating Power	Act 97 (1 period)	• Text 39.3/Lecture	
Radioactive Isotopes		• Text 39.4/Lecture Con Dev Pract Pg 39-1	
Radioactive Half Life		• Text 39.5/Lecture Con Dev Pract Pg 39-2 Activity 98 (<1 period)	
Natural and Artifical Transmutation		• Text 39.6, 39.7/Lecture Con Dev Pract Pgs 39-3, 39-4	
Radioactive Dating		• Text 39.8, 39.9/Lecture	
Radioactive Tracers		• Text 39.10/Lecture	
Radiation and You		• Text 39.11/Lecture	Nx-Time Q 39-2

Video: *Radioactivity*
Evaluation: Chapter 39 Test

See the Chapter Notes for alternative ways to use all these resources

Objectives

After studying Chapter 39, students will be able to:
- Distinguish between the two kinds of nucleons in the nucleus and compare the numbers of each found in the nuclei of different elements.
- Compare the strong force to the electrical force.
- Distinguish among the three types of rays given off by radioactive nuclei and compare their penetrating powers.
- Interpret the symbols used to label isotopes of an element.
- Given the half life of a radioactive isotope and the original amount of the isotope, predict how much of the isotope will remain at the end of some multiple of the half life.
- Given the symbol for a radioactive isotope and the particle it gives off, predict the product of its decay.
- Explain how transuranic elements are produced and why they are not found naturally.
- Give examples of different uses for radioactive isotopes.
- List the major sources of natural background radiation.
- Explain why additional exposure to radiation is harmful.

Possible Misconceptions to Correct

- Radioactivity is sinister, as is everything else we can't see and can't understand.
- Radioactivity is something that has been introduced since 20th-century technology.
- Most of the radiation that people receive stems from 20th-century technology.
- Atoms cannot be changed from one element to another.
- Atoms are the smallest particles of matter that exist.
- Atoms exposed to radiation differ from those that are not exposed to radiation.

No Demonstrations for this Chapter

Introduction

This chapter begins with a description of the atomic nucleus and radioactive decay. Formulas for decay reactions are illustrated with supporting sketches for better comprehension. The background for this material goes back to Chapter 17. This chapter is a prerequisite to Chapter 40.

On text page 614, it is stated that a couple of round-trip flights across this country exposes one to as much radiation as is received in a normal chest X ray. More specifically, a typical dosage of 2 millirems is received when flying across the United States in a jet. This is the same dose received annually from those old luminous dial wristwatches. Cosmic radiation at sea level imparts 45 millirems annually, and radiation from the earth's crust imparts about 80 millirems annually. Living in a concrete or brick house increases this figure slightly for these materials contain more radioactive material than wood does. The human body contains small amounts of carbon-14, potassium-40, and traces of uranium and thorium daughter products, which give one an annual dosage of 25 millirems. So the total natural background radiation annually is about 150 millirems. This makes up about 56% of the radiation the average person encounters, the rest being mainly medical and dental X rays.

Suggested Lecture

If you have discussed the excitation of light previously, consider beginning this lecture by comparing the emission of X rays to the emission of light. Point out that X rays are emitted when the innermost electrons of heavy elements are excited. Cite the fact that the part of the body most prone to radiation damage is the eye — something that dentists should consider taking X rays of the teeth (and inadvertently, the eyes).

Not mentioned in the text is the fact that the discovery of X rays preceded the discovery of radioactivity in 1896 by two months. So the story of radioactivity begins with Wilhelm Roentgen's discovery of X rays.

The Atomic Nucleus: Review the model of the atom with its central nucleus composed of protons and neutrons. Acknowledge that these particles are composed of still smaller particles, quarks, which will not be central to this chapter. Understanding the role of protons and neutrons will be enough. Ask why electrostatic repulsion doesn't make a nucleus fly apart. After all, those protons are very close together! [They don't fly apart because of the action of a stronger force, called the nuclear *strong force*.] Use text Figure 39-3 to explain why big nuclei are unstable.

Radioactive Decay: Distinguish between alpha, beta, and gamma rays. An alpha ray is simply a beam of alpha particles, and a beta ray is a beam of beta particles (electrons). Identify alpha particles as chunks of matter ejected by heavy elements. These ejected chunks are nothing more than the nuclei of helium atoms. The energy they impart to a target is simply their kinetic energy. Once stopped, they are as harmless as cannonballs at rest. Call attention to the fact that the helium commonly used in childrens' balloons is actually the non-radioactive debris of radioactive decay! It is also used as an inert gas to dilute the oxygen content in the tanks of scuba divers. Not all radioactive byproducts are toxic!

If you've covered electricity and magnetism, ask if the rays could be separated by an electric field, rather than by the magnetic field depicted in text Figure 39-5. [Either field will deflect opposite charges in opposite directions.] Compare the penetrating power of these using text Figure 39-8.

Radioactive Isotopes: Distinguish between isotopes and ions (commonly confused), and explain the symbolic way of representing elements (text Figures 39-11, 39-12, and 39-13). Skip over the concept of half life for the time being and demonstrate the symbolic way of writing atomic equations. Write some transmutation formulas on the board while your students follow along with their books opened to the periodic table, text page 247. A repetition and explanation of the reactions shown on text page 606 is in order, if you follow up with one or two new ones as check questions.

For example, have your class write the formula for the alpha decay of Pa-234, which becomes Ac-230, and then for the beta decay of Ac-230, which becomes Th-230.

Beta decay is trickier to understand than alpha decay is, so be sure to treat alpha decay thoroughly. Your students should be able to write their own alpha decay reactions before you write reactions for beta decay.

Treat both natural transmutations and artificial transmutations and put Rutherford's formula, text page 609, on the board. Follow this up with the carbon cycle, text page 610. Now you should discuss half life.

Half Life: Propose this to your students: Jump half way to the wall, then half of the remaining distance, then half of that remaining distance, and so on. How many jumps will get you to the wall? The same situation occurs during radioactive decay. Of course, with a sample of radioactive material, there is a point when all the atoms undergo decay. However, measuring decay rate in terms of this occurrence is a poor idea because of the small sample of atoms present as the process nears the end of its course. The concept of radioactive half life at least insures dealing with half the large number of atoms you begin with.

The concept of half life is established in the exploratory activity with the M and Ms (or coins).

CHECK QUESTION: If the radioactive half life of a certain isotope is one day, how much of the original isotope in a sample will remain at the end of two days? Three days? Four days? [1/4; 1/8; 1/16]

Discuss and compare the various detectors of radiation. If you have the materials for it, demonstrate a cloud chamber.

Cite the role of radioactive isotopes in some kinds of home smoke detectors. These typically use minute amounts of radioactive material, such as americium-241, which is an alpha emitter that transforms air inside the detector chamber into a conductor of electric current. When smoke particles enter the detector, they impede the flow of current through this ionized air and set off an alarm. Thousands of lives are saved each year by these devices — radioactive elements can save lives!

Dating: Return to radioactivity measurement as a means of dating ancient objects by carbon-dating. Discuss the questions posed on text page 611. Cite the usefulness of uranium and other isotope dating in geology.

Radiation and You: Radiation is not good for anybody, but we can't escape it since it is everywhere. However, we can take steps to avoid unnecessary doses. Radiation, like everything that is both damaging and not understood, is usually thought to be worse than it is. You can alleviate a sense of hopelessness by pointing out that it is not new. It is not strictly a product of new technology. As the boy on text page 583 states, the warmth in the inner earth that causes hot springs and geysers is a result of radioactive decay that has been going on since before and during the formation of the earth. It is a part of nature that we must accept. Good sense simply dictates that we avoid unnecessary concentrations of radiation.

More Think-and-Explain Questions

1. Why are alpha and beta particles deflected in opposite directions in a magnetic field? Would they be deflected in opposite directions in an electric field? Why are gamma rays not deflected in either field?

   Answer: Alpha and beta rays are deflected in opposite directions in both a magnetic field and an electric field because they are oppositely charged. Gamma rays have no electric charge and therefore are not deflected.

2. When an alpha particle leaves the nucleus, would you expect it to speed up? Explain.

   Answer: An alpha particle undergoes acceleration due to mutual electric repulsion as soon as it is out of the nucleus and away from the attracting nuclear force. This occurs because the alpha particle has the same charge sign as the nucleus, and like charges repel.

3. If a sample of radioactive material has a half life of 1 week, how much of the original sample will be left at the end of the second week? Third week? Fourth week?

   Answer: After the second week, ¼ will be left; the third week, ⅛; the fourth week, ¹⁄₁₆.

4. A radioisotope placed near a radiation detector registers 80 counts per second. Eight hours later the detector registers 5 counts per second. What is the half life of the radioisotope?

   Answer: Two hours: Since 80 halved 4 times is 5, there have been four half life periods in the 8 hours, and 8 hours/4 = 2 hours.

5. What element results when radium-226 decays by alpha emission? What is the new atomic mass?

   Answer: When radium (atomic number 88) emits an alpha particle, its atomic number is reduced by 2 and it becomes the new element radon (atomic number 86). The resulting atomic mass is reduced by 4 to become radon-226.

40 Nuclear Fission and Fusion

To use this planning guide work from left to right and top to bottom.

Chapter 40 Planning Guide
• The bulleted items are key: Be sure to do them!

Topic	Exploration	Concept Development	Application
Nuclear Fission	Act 99 (<1 period)	• Text 40.1/Lecture	Nx-Time Q 40-1
Fission Reactor		• Text 40.2/Lecture	
Plutonium		• Text 40.3/Lecture	
Breeder Reactor		• Text 40.4/Lecture	
Mass-Energy Relationship		• Text 40.5/Lecture	
Nuclear Fusion		• Text 40.6/Lecture Con Dev Pract Pg 40-1	Nx-Time Q 40-2
Cold Fusion		• Text 40.7/Lecture	
Controlling Nuclear Fusion		• Text 40.8/Lecture	

Video: *Nuclear Fission and Fusion*
Evaluation: Chapter 40 Test

See the Chapter Notes for alternative ways to use all these resources

Objectives

After studying Chapter 40, students will be able to:
- Describe the role of neutrons in causing and sustaining nuclear fission.
- Explain how nuclear fission can be controlled in a reactor.
- Distinguish between a breeder reactor and a uranium-based fission reactor.
- Describe current problems associated with the use of fission as a source of power.
- Predict, from a graph of mass per nucleon vs atomic number, whether energy would be released if a given nucleus split via fission into fragments.
- Distinguish between nuclear fission and nuclear fusion.
- Describe the advantages of fusion over fission as a source of power.
- Describe the current problems associated with using fusion as a source of power.

Possible Misconceptions to Correct

- Nuclear power is sinister, as were electricity, steam power, and other technological advances when they were introduced.
- Nuclear fission and fusion are something new in the universe.
- Plutonium is the most dangerous substance in existence.
- Nuclear fusion can only occur at high temperatures.
- Nuclear power is ecologically more devastating than fossil-fuel power is.

No Demonstration for this Chapter

Introduction

The material in this chapter has a great deal of technological and sociological significance. While nuclear bombs are not avoided in the applications of nuclear energy, the emphasis of the few applications discussed in the chapter is on the positive aspects of nuclear power and its potential for improving the world. Much of the public sentiment against nuclear power has to do with a distrust of what is not understood. Sentiment is against nuclear weapons and nuclear war, rather than with the technological pros and cons of nuclear power. In this climate, we have a responsibility to provide our students with an understanding of the basic physics of nuclear power.

Note that the energy release from the opposite processes of fission and fusion is approached from the viewpoint of decreased mass, rather than the customary treatment of increased binding energy. Hence the usual binding-energy curve is turned upside down in text Figures 40-13, 40-14, and 40-16 to show the relationship of the mass per nucleon versus atomic number. I consider this way conceptually more understandable, for it shows that any reaction involving a decrease in mass releases energy in accordance with mass-energy equivalence.

Mass-energy can be measured in either joules or kilograms (or in ergs or grams). For example, the kinetic energy of a 2-gram beetle walking at the rate of 1 cm/s equals 1 erg, and the energy of the Hiroshima bomb equals 1 gram. So we can express the same quantity using different units.

New to this edition is a section on cold fusion, as muon-induced. This is interesting stuff, and not to be confused with the controversial cold fusion technique presented in 1989 by Professors Fleishman and Pons at the University of Utah. Muon-induced cold fusion is on solid ground, and like other initiators of fusion, energy input exceeds energy output — so far.

A videotape of possible interest to you and your students is *Fusion Torch and Ripe Tomatoes* (45 min). This film presents a speculative and entertaining scenario of a follow-up device to the fusion torch — the replicator, similar to that described by Arthur C. Clark in *Profiles of the Future*.

Suggested Lecture

Nuclear Fission: Review the practice of writing nuclear reactions from the previous chapter and write on the board the fission reaction on text page 619. Discuss the historical significance of the accidental discovery in 1939 in Germany by Otto Hahn and Fritz Strassmann. The discovery was communicated to Lise Meitner and Otto Frisch, then refugees from Nazism in Sweden. Meitner and Frisch recognized its potential and passed the information to American physicists who urged Einstein to write his famous letter urging President Roosevelt to consider the potential use of nuclear fission in warfare. The reaction was considered vitally important, not only because the reaction products had a combined mass less than the mass before reaction which released enormous energy, but also because the reaction released three or so neutrons to produce a chain reaction. This is a good place to consider the dominoes of the exploratory activity, *Chain Reaction*.

Fission Reactors: Explain that ordinary uranium metal doesn't undergo fission because it is mainly composed of the non-fissioning isotope U-238. It is the isotope U-235, which is about 0.7 percent natural uranium, that will spontaneously fission upon neutron capture. These isotopes are "lost" among the more prevalent U-238 or other isotopes. The uranium in reactors is enriched with fissionable isotopes.

Interestingly enough, there is evidence that fission occurred in nature to a small degree millions of years ago when isotopic abundances were different. It occurred in unusually rich concentrations under very unusual circumstances. (*Scientific American*, July 1976.)

Plutonium: Show how U-238 is converted to Pu-239 (see text Figure 40-8). You may wish to discuss the current state of development of fission reactors, particularly breeder reactors.

Mass-Energy Relationship: Compare the masses of different atoms by pretending to grab their nuclei with bare hands and to shake them back and forth. Show with hand motion holding an imaginary giant nucleus how the difference might appear by shaking a hydrogen atom and a lead atom. State that if you were to plot the results of this investigation for all the elements, the relationship between mass and atomic number would look like text Figure 40-12. Draw it on the board. This graph is no big deal. It is not surprising, since atoms of greater atomic number are expected to have greater masses.

Next distinguish between the mass of a nucleus and the mass of the nucleons that make up a nucleus. Ask what a curve of mass/nucleon versus atomic number would look like — that is, if you divided the mass of each nucleus by the number of nucleons composing it and compared the value for different atoms. If all nucleons had the same mass in every atomic configuration, then of course the graph would be a horizontal line, but the masses of nucleons differ. The interrelationship between mass and energy is apparent here, because the nucleons have mass-energy, which is manifested partly in the congealed part (the material matter of the nucleons) and the other part called binding energy. The most energetically-bound nucleus (iron) has the least mass/nucleon. Go into the nucleon-shaking routine again to demonstrate how the nucleons (not the whole nucleus!) become easier to shake as you progress from hydrogen to iron and how they become harder to shake as you progress beyond iron to uranium. Then draw on the board a curve that represents your findings (text Figure 40-13), the most important graph in your course.

From the curve you can show that any nuclear reaction which yields products with less mass than before the reaction occurred will give off energy, and any reaction in which the mass of the product increases will require energy. Further discussion will show how the opposite processes of fission and fusion release energy.

> CHECK QUESTIONS: Which process, fission or fusion, releases energy from atoms of lead? [Fission] Gold? [Fission] Carbon? [Fusion] Neon? [Fusion] Iron? [Neither!] Be careful in selecting atoms too near atomic number 26 in this exercise. For example, elements slightly beyond 26 when fissioned will have more massive products that extend "up the hydrogen hill"; elements near 26 when fused will combine to elements "up the uranium hill." Acknowledging this point, however, may only serve to complicate the picture. Avoid it, unless a student brings it up in class.

Nuclear Fusion: Expand upon the latest developments in *inertial confinement fusion*, which includes fusion induced not only by lasers but also by electron beams and ion beams. Explain how in each case a small fuel pellet is ignited to yield a thermonuclear micro-explosion and how the greatest problem to overcome, other than obtaining significant energies, is the precise timing of laser firings. As of this writing, the Shiva and Novette lasers at Lawrence Livermore Labs have both achieved fusion burns but not sustained burns. Problems have occurred with characteristics of the fuel pellets and the stability of the supporting optics. Sustained fusion by the laser technique is probably many years down the road at best.

Cold Nuclear Fusion: Another exciting entry to the hopes of fusion power is the conceptually simple fusion by muons. When muons take the place of electrons in a hydrogen atom, the electrical barrier is effectively removed. For more on this refer to the article in *Scientific American*, July 1987.

Prospects of Fusion Power: Discussion of the prospects of fusion power is fascinating.

Abundant energy from controlled fusion is one such positive prospect which should concern not only physicists but everyone. Particularly exciting is the prospect of the fusion torch, which may provide a means of recycling material, not to mention the sink it could provide for wastes and pollutants. Ideally, all unwanted waste could be dumped in the fusion torch and vaporized. Atoms could be separated into bins by being beamed through giant mass spectrographs. Point out that the fusion torch may never come to be, because technology may progress further. If the past is any guide, something even better will make this 1970 idea obsolete. Whether or not the fusion torch is imminent, the more important question is to consider how this or comparable achievements will affect the lives of people. How will people interact with one another in a world of relatively abundant energy and material?

More Think-and-Explain Questions

1. Why doesn't uranium ore spontaneously explode?

   Answer: The uranium in the ore is primarily the isotope U-238, which doesn't fission. U-235 atoms in the ore are too far apart for a chain reaction.

2. Which produces more energy — the fissioning of one uranium atom or the fusing of a pair of deuterium atoms? The fissioning of one gram of uranium or the fusing of one gram of deuterium? (Are your answers different? Why?)

   Answer: More energy is released in the fissioning of a single uranium atom than is released in the fusing of a pair of deuterium atoms. However, the much greater number of lighter deuterium atoms in a gram of matter, compared to the fewer heavier uranium atoms in a gram of matter, results in more energy liberated per gram for the fusion of deuterium.

3. Why, unlike fission fuel, is there no limit to the amount of fusion fuel that can be safely stored in one locality?

   Answer: If enough fission fuel is localized, it will ignite spontaneously by the triggering of a single neutron. Fusion fuel, on the other hand, is not ignited this way. It has no critical mass and can be stored in varying amounts without igniting spontaneously.

4. Explain how radioactive decay has always warmed the earth from inside, and nuclear fusion has always warmed the earth from outside.

   Answer: Radioactivity in the earth's core provides heat which keeps the inside molten and warms hot springs and geysers. Nuclear fusion releases energy in the sun, which radiates heat to the earth.

Appendix E: Exponential Growth and Doubling Time

This material is excellent for class discussion. It's not only important, but it is fascinating and very wide in scope. It can be coupled to a discussion of radioactive half life as treated in Chapter 39, or it can be treated in any break. It can follow an exam, or be used on a day that lends itself to a departure from chapter material.

This material is adapted from papers written by Al Bartlett. Consider showing his 50-minute videotape, *The Forgotten Fundamentals of the Energy Crisis*, which comes in three different tape formats for the cost of the tape plus a copying fee. It is available from Academic Media Services, Box 379, University of Colorado, Boulder, CO 80309. For more information on this and other media material by Al Bartlett, call (303) 492 7341.

Exponential doubling time, the time it takes an exponentially-growing quantity to double in size, shows how quickly a quantity can grow to huge proportions. We get the formula for exponential doubling time from the equation

$$N = N_0 e^{kt}$$

Here, N_0 is the initial amount of quantity that increases at a rate k to reach a value of N after time t. Then for doubling time T, $N = 2N_0$, and

$$2N_0 = N_0 e^{kT}$$

Cancelling N_0 and taking the natural logarithm of each side, we get

$$\ln 2 = kT, \text{ or } T = \frac{\ln 2}{k} = \frac{0.693}{k}$$

With k as percent, (so that k = % growth per unit time), $T = 69.3\%/k \cong 70\%/k$. (The same is true for half life, where $N = 0.5N_0$ gives a halving time of $t = \ln 0.5/k. = -0.693/k \cong -70\%/k$. Halving time t is positive since k takes on negative values for decay.)

When percentage figures are given for things such as interest rates, population growth, or the consumption of nonrenewable resources, conversion to doubling time greatly enhances their meaning. For example, saying that the growth rate of a community is 7% annually has little impact. However, saying the growth rate will double the population every ten years is impressive! The following questions can be tackled after your students have read Appendix E.

Questions With Answers

1. For an economy that has a steady inflation rate of 7% per year, how many years would it take for a dollar lose half its value?

   Answer: A dollar loses half its value in one doubling time of the inflationary economy; this is 70/7% = 10 years.

2. At a steady inflation rate of 7%, what will be the price every ten years for the next 50 years of a theater ticket that now costs $10? Of a coat that now costs $100? Of a car that now costs $10 000? Of a home that now costs $100 000?

   Answer: At a steady inflation rate of 7%, the doubling time is 70/7% = 10 years, so every 10 years the prices of these items will double. This means that in 10 years the $10 theater ticket will cost $20; in 20 years, it will cost $40; in 30 years, it will cost $80; in 40 years, it will cost $160; and in 50 years, it will cost $320. Similarly the $100 coat will jump each decade to $200, $400, $800, $1600, and $3200. For a $10 000 car the price jump for each decade would be $20 000, $40 000, $80 000, $160 000, and $320 000. For a $100 000 home, the price jump for each decade would be $200 000, $400 000, $800 000, $1 600 000, and $3 200 000!

3. If the population of a city with one overloaded sewage treatment plant grows steadily at 5% annually, how many overloaded sewage treatment plants will be necessary 42 years later?

   Answer: For a 5% growth rate, 42 years is three doubling times (70/5% = 14 years; 42/14 = 3). Three doubling times is an eight-fold increase. So in 42 years, the city would need 8 sewage treatment plants to remain as overloaded as it is presently, and more than 8 if overloading is to be reduced while servicing 8 times as many people.

4. In 1985 the population growth rate for the United States was 0.7 percent; for Mexico, it was 2.6 percent; and for Kenya with the highest growth rate in the world, it was 4 percent. (This takes into account births, deaths, and immigration.) If these rates were steady, how long would it take for the population in each of these countries to double?

   Answer: For the U.S., 70/0.7 = 100 years; for Mexico, 70/2.6 = 27 years; for Kenya, 70/4 = 17.5 years. (The unusually high growth rates for Mexico and Kenya are typical of underdeveloped countries. They are countries that are the least capable economically of supporting growing populations. In 1986, the average woman in Kenya has eight children.)

5. Suppose that the 4 percent growth rate for the population of Kenya has always been steady and will continue to increase at the same rate for the rest of the century. True or false: In the year 2002 there will be more Kenyan teenagers than the total number of Kenyan people who ever lived.

Answer: That is true for children seventeen years and younger. If you count 18- and 19-year-olds, the number of teens *before* the year 2000 in Kenya will outnumber the total number of Kenyans who ever lived. This assumes that growth will continue to be a steady 4 percent annually. (In the last decade, the increase in world population is more than the population of India. Can you understand better the statement made by Professor Bartlett and found in the footnote on text page 646?)

6. If the world's population doubles in 40 years and the world's food production also doubles in 40 years, how many people will be starving each year compared to now?

Answer: All things being equal, the doubling of food production for twice the number of people simply means that twice as many people will be eating and twice as many will be starving.

7. Suppose you get a prospective employer to agree to hire your services for a wage of a single penny for the first day, 2 pennies the second day, and doubling each day's wage thereafter (providing the employer keep to the agreement for a month). What will be your total wages for the month?

Answer: Doubling one penny for 30 days yields a total of $10 737 418.23.

8. In the last question, how will your wages be for only the thirtieth day compared to your total wages for the previous 29 days?

Answer: On the thirtieth day your wages will be $5 368 709.12, which is one penny more than the $5 368 709.11 total from all the preceding days.

9. Oil has been produced in the U.S. for about 120 years. If there is an equal amount of oil left to be discovered, what is wrong with the argument that the remaining oil will be sufficient for another 120 years?

Answer: The argument that half our oil reserves have served us for 120 years, and that the remaining half therefore ought to serve us for another 120 years fails to take into account the growth in consumption. If consumption grows at a steady rate then the remaining oil will be used in one doubling time, not in 120 years.

10. If fusion power were harnessed today, the resulting abundant energy is likely to sustain and to encourage even further our present appetite for continued growth. Also, in a relatively few doubling times, it would produce an appreciable fraction of the solar power input to the earth. Make an argument that the current delay in harnessing fusion is a blessing for the human race.

Answer: It is generally acknowledged that if the human race is to survive, even from a standpoint of overheating of the world, the present rate of energy consumption and population growth must be reduced. The chances of achieving a reduced growth rate are better in a climate of scarce energy than in a climate of abundant energy. We must hope that by the time we have fusion under control, we will have learned to optimize our numbers and to use energy more wisely.